MODALITIES

MODALITIES

Philosophical Essays

RUTH BARCAN MARCUS

New York   Oxford
OXFORD UNIVERSITY PRESS

Oxford University Press

Oxford   New York   Toronto
Delhi   Bombay   Calcutta   Madras   Karachi
Kuala Lumpur   Singapore   Hong Kong   Tokyo
Nairobi   Dar es Salaam   Cape Town
Melbourne   Auckland

and associated companies in
Berlin   Ibadan

Copyright © 1993 by Ruth Barcan Marcus

First published in 1993 by Oxford University Press, Inc.
200 Madison Avenue, New York, NY 10016
First issued as an Oxford University Press paperback, 1995

Oxford is a registered trademark of Oxford University Press

Library of Congress Cataloging-in-Publication Data
Marcus, Ruth Barcan.
Modalities : philosophical essays / Ruth Barcan Marcus.
p. cm.   Includes bibliographical references.
ISBN 0-19-507698-2; 0-19-509657-6 (pbk)
1. Metaphysics.
2. Ontology.
3. Language and logic.
4. Modality (Logic)
I. Title.
BD111.M44   1993   110—dc20
91-48105

9 8 7 6 5 4 3 2

Printed in the United States of America
on acid free paper

For Jim, Peter, Katherine, and Libby

Contents

Acknowledgments

The present volume contains a selection of essays written during the past three decades. There is a roster of many to whom I owe thanks for encouraging the logical and philosophical interests that are reflected in these papers.

While I was an undergraduate, J. C. C. McKinsey supervised my study of mathematical logic, spending countless hours with me as a self-elected informal tutor. I took for granted what in retrospect I see as extraordinary generosity. McKinsey also encouraged my more specialized interest in modal logic and in the philosophical questions that related to those interests. Curricular efforts were directed toward gaining a "classical" education as viewed by a romantic in her teens. In addition to philosophy and mathematics, I studied some physics, some history and classical languages. Little remains of the latter. My literary preoccupations were extracurricular, for I had located myself in a student fringe of the *Partisan Review* milieu.

McKinsey encouraged me to go on to graduate work at Yale, where my already germinated work in modal theories would be taken seriously. At Yale five (or was it six?) marathon examinations in traditional fields of Philosophy were required. Beyond that, I was left to my own devices. In addition to some philosophy seminars I struggled through a year-long beautifully structured six-day-a-week course in theoretical physics taught by Leigh Page and required of first-year physics graduate students. I also audited mathematics lectures of Einar Hille and Oystein Ore. Those disciplines helped shape my philosophical thought.

Yale philosophy faculty members who were particularly influential at that time were Charles Stevenson and Ernst Cassirer. Cassirer was a profound philosopher and a dazzling polymath who succeeded in diverting my Vienna Circle predilections into some Kantian channels. Above all I worked with Frederic B. Fitch, who endorsed my proposal to develop quantified modal logic as a dissertation. My thesis was written in absentia and sent piecemeal for approval until (since I seemed to be going on indefinitely), Fitch ruled that enough was enough. Fitch's low-key support and encouragement were unstinting.

The dissertation was completed in early 1945 except for a final typing, which I did sporadically on my return to New Haven.

During 1945–1947 I was engaged as a research associate on a cross-cultural anthropological study grounded in psychoanalytic as well as behavioral theory. The outcome surfaced as a book by J. W. M. Whiting and Irvin Child, *Child Training and Personality* (Yale University Press, 1953) in which I am surprisingly cited as "an important contributor and critic of concepts in the initial planning of the research." I learned much, and John Whiting's brainstorming style has left its mark.

During 1946 and 1947 some of my work in quantified modal logic appeared in the *Journal of Symbolic Logic* and almost immediately thereafter was reviewed in that journal by W. V. Quine. His reviews voiced doubts about possible interpretations, doubts that were also expressed in "The Problem of Interpreting Modal Logic" (ibid., 1947). Those criticisms and the continuing debate were a catalyst for some of my subsequent work.

In 1947–1948 a postdoctoral fellowship took me to the University of Chicago, where I joined Rudolf Carnap's seminar. Among the participants were William Alston, Richard Jeffreys, and Howard Stein. Contrary to rumor, Carnap was wholly undogmatic and created an atmosphere of mutual respect in the cooperative enterprise of philosophical inquiry. For many years thereafter and without regular affiliation with a major department I continued to work autonomously while resident in the Chicago area. I was encouraged by increased attention to my published papers.

During 1961 I wrote "Modalities and Intensional Languages," which was presented to the Boston Colloquium. It is the first paper selected for this collection. Saul Kripke was present on that occasion, and his support was warmly appreciated. I was then teaching part-time at Roosevelt University. In 1962 Arthur Prior visited the University of Chicago and I attended his seminars when I could. We were kindred spirits. It was at his urging that I participated in the Helsinki conference on modal and many-valued logics during the summer of 1962. That conference marked a renaissance of interest in intensional and nonstandard formal theories.

In the mid-1960s I was entrusted with the task of constituting a philosophy department at the newly established Chicago campus of the University of Illinois. It was an onerous though exciting undertaking and the faculty assembled were wonderful colleagues. From there I moved to Northwestern and then to Yale. My interest in intensional theories continued, but my work broadened over the years to include topics in metaphysics and epistemology as well as moral philosophy.

The covering notes to the essays as well as the footnotes indicate who are my intellectual creditors. But there are those whose encouragement and interest also helped me (perhaps unbeknownst to some of them) in less direct but significant ways. In addition to those mentioned above, they are Joseph Almog, G. E. M. Anscombe, Alan Anderson, Nuel Belnap, Paul Benacerraf, Sarah Waterlow Broadie, Nancy Cartwright, Vere Chappell, Alonzo Church, George Dickie, Joel Feinberg, Philippa Foot, Robert Fogelin, Dagfinn Føllesdal, Adolf Grünbaum, David Kaplan (a major influence and continuing inspiration), Raymond Klibansky, Leonard Linsky, Jane Isay, Mary Mothersill, John Perry, Charles Parsons, Sir Peter Strawson, Bryan Skyrms, Nathan Salmon, Scott Soames, William Tait, Judith Thomson and Howard Wettstein. There are, I expect, omissions that will return to haunt me.

My thanks also to the students I have had whose philosophical enthusiasms I shared. Many of them, graduate and undergraduate, have gone on to distinguished careers in philosophy, which is a source of considerable pleasure.

I have been supported intermittently since 1953 by the Guggenheim Foundation, the National Science Foundation, residencies at the Rockefeller Foundation Center in Bellagio, the Center for Advanced Study in the Behavioral Sciences at Stanford, and visiting fellowships at the Edinburgh Institute for Advanced Study, Wolfson College in Oxford, and Clare Hall in Cambridge. The Philosophy Department at the University of California in Los Angeles graciously received me as a visitor during a Yale leave.

Leigh Cauman read the essays, and her editorial wisdom has improved them. At Oxford University Press, Cynthia Read and Melinda Wirkus patiently steered the book to final publication. The Yale Griswold Fund partially supported preparation of the manuscript by Glena Ames, whose secretarial assistance was virtually flawless.

Permission has been granted where required by editors and publishers for reprinting of papers in the present volume.

My thanks to all.

New Haven   R. B. M.
September 1991

Introduction

The papers included here were written as freestanding essays. Central themes of each often rely on previous work, and there is consequently too much repetition. The alternative was radical editing, which would have made the volume questionable as a collection of previously published essays.

The essays are presented chronologically, beginning with "Modalities and Intensional Languages" (1961–1962). Several themes introduced there thread their way through many of the papers concerned with logical, semantical, metaphysical, and epistemological issues in intensional logic and, in particular, modalities. They concern the following:

Extensionality. Extensionality is not taken as a single principle. It is defined as characterizing a language or a theory, where a language or a theory is regarded as extensional relative to the extent to which stronger equivalence relations are equated with weaker ones. This may be done explicitly, as, for example, where property identity is reduced to set identity, or implicitly by restricting permissible predicates or contexts in such a way as to achieve a reduction that supports intersubstitutivity.

The necessity of identity. The formal proof of the necessity of identity for systems of second-order quantified modal logic is in one of several technical papers published in 1946–1947 in the *Journal of Symbolic Logic*. It should be noted that the proof does not depend on any peculiarly modal postulates. The necessity of identity is represented in various philosophical contexts in several subsequent papers here included. In particular, in my papers on classes, attributes, and collections (one of which, titled "Classes, Collections, Assortments, and Individuals," is included here) necessary identity for collections is a derivative extension of necessary identity for individuals.

Proper names as directly referential tags. A sharp semantical distinction is drawn between names and descriptions. As in the initial essay, "a proper name [of a thing] has no meaning. It is not strongly equatable with any of the singular descriptions of the thing. . . ." Proper names are not assimilated to what later came to be called "rigid designators'' by Saul Kripke, although they share some features with rigid descriptions. The view of proper names as tags along with the provable constraints on substitutivity of weak equivalents (under which the theory of descriptions may be viewed as falling) dispels ''puzzles'' about substitution in modal and, arguably, other intensional contexts.

Essentialism. Accounts of essential properties are considered that accommodate an Aristotelian notion of essential property. On those accounts, it is argued that modal languages need not commit one to essentialism. It is *not* argued that essentialism is invidious. Essentiality for certain properties (e.g., kinds) is defended.

Substitutional quantification. A substitutional account of the quantifiers is considered as a possible alternative to standard semantics for quantification theory for intensional languages. Some limitations on the substitutional interpretation are noted where the identity relation for individuals is introduced. Although substitutional quantification is viewed as plausible and defensible, the inital essay presents in conclusion an objectual interpretation that supports the Barcan formula $[\Diamond(\exists x)A \dashv 3 \ (\exists x)\Diamond A]$.

Possibilia and possible worlds. My interpretation of modal theories is grounded in the supposition of actualism. There are puzzles about possibilia (nonactual individual objects), and arguments for ''their'' exclusion are given.

Included here are two papers on belief that have some consequences for epistemic logic but that go far beyond those considerations. They have, *inter alia,* some import for theories of rationality. Also included are two papers that have some consequences for deontic logic and practical reasoning but that are addressed to wider ethical issues. In particular, in ''Moral Dilemmas and Consistency,'' it is argued that moral dilemmas are not evidence of inconsistency of a moral code and are not evidence for antirealistic views.

I have also included two papers addressed to particular historical figures, Russell and Spinoza. And, of course, Quine's despair of intensional languages is a counterpoint to many of the essays.

MODALITIES

1. Modalities and Intensional Languages

The source of this paper was published in *Synthese,* XIII, 4 (December 1961): 303–322. It was presented in February 1962 at the Boston Colloquium for the Philosophy of Science in conjunction with a commentary by W. V. Quine. Quine's comments were also published in *Synthese*, XIII, and later republished under the title "Reply to Prof. Marcus." My paper, the comments, and a discussion in *Synthese* XIV listing R. B. Marcus, W. V. Quine, S. Kripke, J. McCarthy, and D. Føllesdal as discussants also appeared in *Boston Studies in the Philosophy of Science,* ed. Marx Wartofsky (Dordrecht: Reidel, 1963). The discussion is included here as an appendix. The present printing of my paper contains some editorial corrections and nonsubstantive changes, some of which were included in the reprinting in *Contemporary Readings in Logical Theory,* ed. I. Copi and J. Gould (New York: Macmillan, 1967).

I was especially appreciative of the presence of Saul Kripke and of his participation in the discussion at the Boston Colloquium since Kripke was clearly open to my interests at a time when concern with modal issues was viewed askance. It is worth noting that in saying, in the text that follows, that "I have never appreciated the force of the original argument" about failures of substitutivity in modal contexts, I had assumed that Arthur Smullyan's paper "Modality and Description," *Journal of Symbolic Logic,* XIII (1948): 31–37, which I reviewed in the same journal during that year, pp. 149–150, was fully appreciated. Smullyan had shown that Russell's theory of descriptions, properly employed with attention to scope, dispelled the puzzles. Smullyan's solution was also elaborated and extended by F. B. Fitch in "The Problem of the Morning Star and the Evening Star," *Philosophy of Science,* XVI (1949). However, the absence of appreciation of Smullyan's account persisted, and I have therefore included my review of his paper as a second appendix.

It was not unusual at the time of presentation of this paper, even among the most rigorous of logicians, to use 'tautology' and its cognates for 'logically valid' or 'valid' and sometimes for 'analytic'; similarly, 'value of a variable' was used as an alternative to 'substituend of a variable'. 'Implication' was sometimes used for 'consequence' and sometimes for 'conditional'. Such alternative usage did not always signify use/mention confusions. I have made a few replacements in this volume, consistent with present more standardized practices but only if such replacements made no substantive difference. ■

There is a normative sense in which it has been claimed that modal logic is without foundation. W. V. Quine, in *Word and Object,* seems to believe that it was conceived in sin: the sin of confusing use and mention. The original transgressors were Russell and Whitehead. C. I. Lewis followed suit and constructed a logic in which an operator corresponding to 'necessarily' operates on sentences, whereas 'is necessary' ought to be viewed as a predicate of sentences. As Quine reconstructs the history of the enterprise,[1] the operational use of modalities promised only one advantage: the possibility of quantifying into modal contexts. This some of us[2] were enticed into doing. But the evils of the modal sentential calculus were found out in the functional calculus, and with them—to quote from *Word and Object*—"the varied sorrows of modality."

I do not intend to claim that modal logic is wholly without sorrows. I do claim that its sorrows are not those that Quine describes, and that modal logic is worthy of defense, for it is useful in connection with many interesting and important questions, such as the analysis of causation, entailment, obligation, and belief statements, to name only a few.

If we insist on equating formal logic with strongly extensional functional calculi, then P. F. Strawson[3] is correct in saying that "the analytical equipment (of the formal logician) is inadequate for the dissection of most ordinary types of empirical statement."

Intensional Languages

I will begin with the notion of an intensional language. I will make a further distinction between those languages that are explicitly intensional and those that are implicitly so. Our notion of intensionality does not divide languages into mutually exclusive classes but, rather, orders them loosely as strongly or weakly intensional. A language is explicitly intensional to the degree to which it does not equate the identity relation with some weaker form of equivalence. We will as-

---

1. *Word and Object* (Cambridge: Harvard University Press, 1960), pp. 195–96.
2. a. R. C. Barcan (Marcus), "A Functional Calculus of First Order Based on Strict Implication," *Journal of Symbolic Logic,* XI (1946): 1–16.
   b. R. C. Barcan (Marcus), "The Identity of Individuals in a Strict Functional Calculus of First Order," *Journal of Symbolic Logic,* XII (1947): 12–15.
   c. R. Carnap, "Modalities and Quantification," *Journal of Symbolic Logic,* XI (1946): 33–64.
   d. F. B. Fitch, *Symbolic Logic* (New York: Ronald Press, 1952).
3. P. F. Strawson, *Introduction to Logical Theory* (London: Methuen, 1952), p. 216.

sume that every language must have some constant objects of reference (things), ways of classifying and ordering them, ways of making statements, and ways of separating true statements from false ones. We will not go into the question of how we come to regard some elements of experience as things, but one criterion for sorting out the elements of experience that we regard as things is that things may enter into the identity relation. In a formalized language, those symbols that name things will be those for which it is meaningful to assert that 'I' names the identity relation.

Ordinarily, and in the familiar constructions of formal systems, the identity relation must be held appropriate for individuals. If 'x' and 'y' are individual names, then

(1) $x\mathrm{I}y$

is a sentence, and, if they are individual variables, then (1) is a sentential function. Whether a language confers thinghood on attributes, classes, or propositions is not so much a matter of whether variables appropriate to them can be quantified over (and we will return to this later), but rather of whether (1) is meaningful where 'x' and 'y' may take as substitution instances names of attributes, classes, propositions. We note in passing that the meaningfulness of (1) with respect to attributes and classes is more frequently allowed than the meaningfulness of (1) in connection with propositions.

Returning now to the notion of explicit intensionality, if identity is appropriate to propositions, attributes, classes, as well as lowest-type individuals, then any weakening of the identity relation with respect to any of these entities may be thought of as an extensionalizing of the language. By a weakening of the identity relation is meant equating it with some weaker equivalence relation.

On the level of individuals, one or perhaps two equivalence relations are customarily present: identity and indiscernibility. This does not preclude the introduction of others, such as congruence, but the strongest of these is identity. Where identity is defined rather than taken as primitive, it is customary to define it in terms of indiscernibility. Indiscernibility may in turn be defined as

(2) $x\ \mathrm{Ind}\ y =_{\mathrm{df}} (\varphi)(\varphi x\ \mathrm{eq}\ \varphi y)$

for some equivalence relation eq.

In a system of material implication (Sm), eq is taken as $\equiv$ (material equivalence). In modal systems, eq may be taken as $\equiv$ (strict equivalence). In more strongly intensional systems eq may be taken as the strongest equivalence relation appropriate to such expressions as 'φx'.

In separating (1) and (2) I should like to suggest the possibility that to equate identity and Ind may already be an explicit weakening of the identity relation and, consequently, an extensionalizing principle. This was first suggested to me by a paper of F. P. Ramsey.[4] Though I now regard his particular argument in support of the distinction as unconvincing, I am reluctant to reject the possibility. I suppose that at bottom my appeal is to ordinary language, since, although it is obviously absurd to talk of two things being the same thing, it seems not quite so absurd to talk of two things being indiscernible from each other. In equating I and Ind we are saying that to be distinct is to be discernibly distinct in the sense of there being one property not common to both. Perhaps it is unnecessary to mention that, if we confine things to objects with spatiotemporal histories, it makes no sense to distinguish (1) and (2). And indeed, in my extensions of modal logic, I have chosen to define identity in terms of (2). However, the possibility of such a distinction ought to be mentioned before it is obliterated. Except for the weakening of (1) by equating it with (2), extensionality principles are absent on the level of individuals.

Proceeding now to functional calculi with theory of types, an extensionality principle is of the form

(3) $x \text{ eq } y \rightarrow x I y$

The arrow may represent one of the implication relations present in the system or some metalinguistic conditional; eq is one of the equivalence relations appropriate to x and y, but not identity. Within the system of material implication, 'x' and 'y' may be taken as symbols for classes, 'eq' as class equality (in the sense of having the same members); or 'x' and 'y' may be taken as symbols for propositions and 'eq' as the triple bar for material equivalence. In extended modal systems 'eq' may be taken as the quadruple bar where 'x' and 'y' are symbols for propositions. If the extended modal system has symbols for classes, 'eq' may be taken as 'having the same members' or, alternatively, 'necessarily having the same members', which can be expressed within such a language. If we wish to distinguish classes from attributes in such a system 'eq' may be taken as 'necessarily applies to the same thing', which is directly expressible within the system. In a language that permits epistemic contexts such as belief contexts, an equivalence relation even stronger than either material or strict equivalence may have to be present. Taking that stronger relation as

---

4. F. P. Ramsey, *The Foundations of Mathematics* (London and New York: Methuen, 1931), pp. 30–32.

eq, (3) would still be an extensionalizing principle in such a strongly intensional language.

I should now like to turn to the notion of implicit extensionality, which is bound up with the kinds of substitution principles available in a language. If we confine ourselves for the sake of simplicity of exposition to a sentential calculus, one form of the substitution theorem is

(4) x eq$_1$ $y \rightarrow z$ eq$_2$ w

where x, y, z, w are well formed, w is the result of replacing one or more occurrences of x by y in z, and '$\rightarrow$' symbolizes an implication or a metalinguistic conditional. In the system of material implication (Sm or QSm), (4) is provable where eq$_1$ and eq$_2$ are both taken as material equivalence for appropriate values of x, y, z, w. That is,

(5) $(x \equiv y) \supset (z \equiv w)$

Now (5) is clearly false if we are going to allow contexts involving belief, logical necessity, physical necessity, and so on. We are familiar with the examples. If 'x' is taken as 'John is a featherless biped', and 'y' as 'John is a rational animal', then an unrestricted (5) fails. Our choice is either to reject (5) as it stands or to reject all contexts in which it fails. If the latter choice is made, the language is *implicitly* extensional, since it cannot countenance predicates or contexts that might be permissible in a more strongly intensional language. This is Quine's solution. All such contexts are assigned to a shelf labeled 'referential opacity' or, more precisely, 'contexts that confer referential opacity', and are disposed of. But the contents of that shelf are of enormous interest to some of us, and we would like to examine them in a systematic and formal manner. For this we need a language that is appropriately intensional.

In the modal sentential calculi, since there are two kinds of equivalence that may hold between x and y, (4) represents several possible substitution theorems, some of which are provable. We will return to this shortly. Similarly, if we are going to permit epistemic contexts, the modal analogue of (4) will fail in those contexts and a more appropriate principle will have to supplement it. A stronger eq$_1$ is required to support substitutivity in epistemic contexts.

Identity and Substitution in Quantified Modal Logic

In the light of the previous remarks I would like to turn specifically to the criticisms raised against extended modal systems in connection with identity and substitution. In particular, I will refer to my[5] extension of Lewis's[6] S4, which consisted of introducing quantification in the usual manner and adding the axiom[7]

(6) $\Diamond(\exists x)A \dashv 3 (\exists x)\Diamond A$

I will call this system 'QS4'. QS4 does not have an explicit axiom of extensionality, although it does have an implicit weak extensionalizing principle in the form of a substitution theorem.

It would appear that, for many uses to which modal calculi may be put, S5 is to be preferred.[8] A. N. Prior[9] has shown that (6) is a theorem in an extended S4 (i.e., S5). My subsequent remarks, unless otherwise indicated, apply equally to QS5. In QS4 (1) is defined in terms of (2). (2), and consequently (1), admit of alternatives where 'eq' may be taken as material or strict equivalence: 'I_m' and 'I' respectively. The following are theorems of QS4:

(7) $(xI_m y) \equiv (xIy)$

(8) $(xIy) \equiv \Box(xIy)$

where '$\Box$' is the modal symbol for logical necessity. In (7) 'I_m' and 'I' are strictly equivalent; within such a modal language, they are therefore indistinguishable by virtue of the substitution theorem. Contingent identities are disallowed by (8).

(9) $(xIy) \cdot \Diamond \sim(xIy)$

is a contradiction.

Professor Quine[10] finds these results offensive, for he sees (8) as "purifying the universe." Concrete entities are said to be banished and replaced by pallid concepts. The argument is familiar:

(10)   The evening star eq the morning star

5. Barcan (Marcus), "A Functional Calculus of First Order Based on Strict Implication"; "The Identity of Individuals in a Strict Functional Calculus of First Order."

6. C. I. Lewis and C. H. Langford, *Symbolic Logic* (New York: Methuen, 1932).

7. See A. N. Prior, *Time and Modality* (New York: Oxford University Press, 1957), for an extended discussion of this controversial axiom, which has come to be known as the Barcan formula.

8. S5 results from adding to S4:

   $p \dashv 3 \Box \Diamond p$

9. A. N. Prior, "Modality and Quantification in S5," *Journal of Symbolic Logic*, XXI (1956), pp. 60–62.

10. *From a Logical Point of View* (Cambridge: Harvard University Press, 1953), pp. 152–154.

is said to express a "true identity," yet the expressions on either side of 'eq' are not validly intersubstitutable in

(11)   It is necessary that the evening star is the evening star.

The rebuttals are familiar, and I will try to state some of them. This is difficult, for I have never appreciated the force of the original argument. In restating the case, I would like to consider the following informal claim:

(12)   If p is a tautology, and p eq q, then q is a tautology.

where 'eq' names some equivalence relation appropriate to p and q. In Sm, if 'eq' is taken as $\equiv$, then a restricted (12) is available where '$p \equiv q$' is provable.

One might say informally that, with respect to any language, if (12) is said to fail, then we must be using 'tautology' in a very peculiar way, or what is taken as 'eq' is not sufficient as an equivalence relation appropriate to p and q.

Consider by contrast the claim that

(13) $a\mathrm{I}b$

is a true identity. Now if (13) is such a true identity, then a and b are the same thing. (13) doesn't say that a and b are two things that happen, through some accident, to be one. True, we are using two different names for that same thing, but we must be careful about use and mention. If, then, (13) is true, it must say the same thing as

(14) $a\mathrm{I}a$

But (14) is surely valid, and so (13) must surely be valid as well. This is precisely the import of my theorem (8). We would therefore expect, indeed it would be a consequence of the truth of (13), that 'a' is replaceable by 'b' in any context.

Now suppose we come upon a statement like

(15)   Scott is the author of *Waverley*.

and we have a decision to make. This decision cannot be made in a formal vacuum, but must depend to a considerable extent on some informal consideration as to what it is we are trying to say in (10) and (15). If we decide that 'the evening star' and 'the morning star' are proper names for the same thing, and that 'Scott' and 'the author of *Waverley*' are proper names for the same thing, then they must be intersubstitutable in every context. In fact it often happens, in a growing, changing language, that a descriptive phrase comes to be used as a proper name—an identifying tag—and the descriptive meaning is lost or ignored. Sometimes we use certain devices, such as capitalization with or without dropping of the definite article, to indicate the change in use. 'The evening star' becomes 'Evening Star', 'the morning star' becomes 'Morning Star', and they may come to be used as names for the same thing. Singular descriptions such as 'the little corporal', 'the Prince of Denmark', 'the sage of Concord', or 'the great dissenter', are, as we know, often used as alternative proper names of Napoleon, Hamlet, Emerson, and Oliver Wendell Holmes. One might even devise a criterion as to when a descriptive phrase is being used as a proper name. Suppose that, through some astronomical cataclysm, Venus was no longer the first star of the evening. If we continued to call it alternatively 'Evening Star' or 'the evening star' that practice would be a measure of the conversion of the descriptive phrase into a proper name. If, however, we would then regard (10) as false, this would indicate that 'the evening star' was not being used as an alternative proper name of Venus. (We might mention in passing that, although the conversion of descriptions into proper names appears to be asymmetric, we do find proper names used in singular descriptions of something other than the thing named, as in the statement 'Mao Tse-tung is the Stalin of China', where one intends to assert a similarity between the entities named.)

That any language must countenance some entities as things would appear to be a precondition for language. But this is not to say that experience is given to us as a collection of things, for it would appear that there are cultural variations and accompanying linguistic variations as to what sorts of entities are so singled out. It would also appear to be a precondition of language that the singling out of an entity as a thing is accompanied by many—and perhaps an indefinite number— of unique descriptions, for otherwise how would it be singled out? But to assign a thing a proper name is different from giving a unique description. For suppose we took an inventory of all the entities countenanced as things by some particular culture through its own language, with its own set of names and equatable singular descriptions, and suppose that number were finite (this assumption is for the sake of simplifying the exposition). And suppose we randomized as many whole numbers as we needed for a one-to-one correspondence, and thereby tagged each thing. This identifying tag is a proper name of the thing. In taking our inventory we discovered that many of the entities countenanced as things by that language-culture complex already had proper names, although in many cases a singular description may have been used. This tag, a proper name, has no meaning. It simply tags. It is not strongly equatable with any of the singular descriptions of the thing, although singular descriptions may be equatable (in a weaker sense) with each other, where

(16) $Desc_1$ eq $Desc_2$

means that $Desc_1$ and $Desc_2$ describe the same thing. But here too, what we are asserting would depend on our choice of 'eq'.

Perhaps I should mention that I am not unaware of the awful simplicity of the tagging procedure described above. It makes the assumption of finitude, or, if this is not assumed, then at least the assumption of denumerability of the class of things; also, the assumption that all things countenanced by the language-culture complex are named (or nameable), described (or describable). But my point is only to distinguish tagging from describing, proper names from descriptions. You may describe Venus as the evening star, and I may describe Venus as the morning star, and we may both be surprised that, as an empirical fact, the same thing is being described. But it is not an empirical fact that

(17)   Venus I Venus

and if 'a' is another proper name for Venus that

(18)   Venus I a

Nor is it extraordinary that we often convert one of the descriptions of a thing into a proper name. Perhaps we ought to be more consistent in our use of upper-case letters, but this is a question of reforming ordinary language. It ought not to be an insurmountable problem for logicians.

What I have been arguing is that to say truly of an identity (in the strongest sense of the word) that it is true, it must be tautologically true or analytically true. The controversial (8) of QS4, the necessity of identity, no more banishes concrete entities from the universe than (12) banishes from the universe red-blooded propositions.

Let us now return to (15) and (10). If they express true identities, then 'Scott' ought to be anywhere intersubstitutable for 'the author of *Waverley*' in modal contexts, and similarly for 'the morning star' and 'the evening star'. If those pairs are not so universally intersubstitutable—that is, if our decision is that they are not simply proper names for the same thing; that (15) and (10) express equivalences that are possibly false, e.g., someone else might have written *Waverley*, the star first seen in the evening might have been different from the star first seen in the morning—then (15) and (10) do not express identities.

Russell provides one solution; on his analysis in accordance with the theory of descriptions the truth of (15) and (10) does not commit us to their necessary truth, and certainly not to taking 'eq' of (10) or 'is' of (15) as identity, except on the explicit assumption of an extensionalizing axiom. Other and related solutions are in terms of membership in a unit class or applicability of a unit attribute. But, whatever we choose, it will have to permit intersubstitutability or some analogue of intersubstitutability for the members of the pairs 'Scott' and 'the author of *Waverley*', and 'the evening star' and 'the morning star', which is short of being universal. In a language that is implicitly strongly extensional, that is, where all contexts in which such substitutions fail are simply eschewed, of course there is no harm in equating identity with weaker forms of equivalence. But why restrict ourselves in this way when, in a more intensional language, we can still make all the substitutions permissible to this weaker form of equivalence, yet admit contexts in which such substitution is not permitted? To illustrate, I would like to turn to the instances of (4) that are provable[11] in QS4. I will again confine my remarks, for the purpose of exposition, to S4, although it is the generalizations for QS4 that are actually proved. An unrestricted

(19) $\quad x \equiv y \rightarrow z \equiv w$

is not provable, whether '$\rightarrow$' is taken as material implication, strict implication, or a metalinguistic conditional. It would involve us in a contradiction if our interpreted system allowed true statements such as

(20) $\quad (x \equiv y) \cdot \sim \Box(x \equiv y)$

as it must if it is not to reduce itself to the system of material implication. Indeed, the underlying assumption about equivalence that is implicit in the whole "evening star/morning star" controversy is that there are equivalences (misleadingly called "true identities") that are contingently true. Let x and y of (19) be taken as some pair p and q that satisfies (20). Let z be $\Box(p \equiv p)$ and w be $\Box(p \equiv q)$. Then (19) is

(21) $\quad (p \equiv q) \rightarrow (\Box(p \equiv p) \equiv \Box(p \equiv q))$

11. Barcan (Marcus), "A Functional Calculus of First Order Based on Strict Implication." Theorem XIX* corresponds to (23). The restricted (19), given the conditions of the restriction, is clearly provable in the same manner as XIX*.

From (20), simplification, *modus ponens,* and $\Box(p\equiv p)$, which is a theorem of S4, we can deduce $\Box(p\equiv q)$. $\Box(p\equiv q)$, simplification of (20), and conjunction lead to the contradiction

(22) $\quad \Box(p\equiv q)\cdot\sim\Box(p\equiv q)$

A restricted form of (19) is provable: (19) is provable where z does not contain any modal operators. And these are the contexts that support substitution in Sm (the system of material implication), without at the same time banishing modal contexts. Indeed, a slightly stronger (19) is provable. (19) is provable where x does not fall within the scope of a modal operator in z.

Where in (4), eq_1 and eq_2 are both taken as strict equivalence, the substitution theorem

(23) $\quad (x\equiv y)\rightarrow(z\equiv w)$

is provable without restriction, and also where eq_1 is taken as strict equivalence and eq_2 is taken as material equivalence, as in

(24) $\quad (x\equiv y)\rightarrow(z\equiv w)$

But (23) is also an extensionalizing principle, for it fails in epistemic contexts such as contexts involving 'knows that' or 'believes that', for consider the statement

(25)   When Professor Quine reviewed the paper on identity in QS4, he knew that $\vdash a\mathrm{i}_m b\equiv a\mathrm{I}_m b$.

and

(26)   When Professor Quine reviewed the paper on identity in QS4 he knew that $\vdash a\mathrm{I}b\equiv a\mathrm{I}_m b$.

Although (25) is true, (26) is false, even though (7) holds in QS4. But rather than repeat the old mistakes by abandoning epistemic contexts to the shelf labeled "referential opacity" after having rescued modal contexts as the most intensional permissible contexts to which such a language is appropriate, we need only conclude that (23) confines us to limits of applicability of such modal systems. *If* it should turn out that statements involving 'knows that' and 'believes that' permit of formal analysis, then such an analysis would have to be embedded in a language with a stronger equivalence relation for sentences than strict equivalence. Carnap's intensional isomorphism, Lewis's analytical comparability, and perhaps Anderson and Belnap's mutual entailment are relations that point in that direction. But they too would be short of identity, for there surely are contexts in which substitutions allowed by such stronger equivalences would convert a truth into a falsehood.

It is my opinion [12] that the identity relation need not be introduced for anything other than the entities we countenance as things, such as individuals. Increasingly strong substitution theorems give the force of universal substitutivity without explicit axioms of extensionality. We can talk of equivalence between propositions, classes, attributes, without thereby conferring on them thinghood by equating such equivalences with the identity relation. QS4 has no explicit extensionality axiom. Instead we have (23), the restricted (19), and their analogues for attributes or classes.

The discussion of identity and substitution in QS4 would be incomplete without touching on the other familiar example:

(27)   9 eq the number of planets

is said to be a true identity for which substitution fails in

(28) $\Box(9 > 7)$

for it leads to the falsehood

(29) $\Box$(the number of planets > 7)

Since the argument holds (27) to be contingent [$\sim\Box$(9 eq the number of planets)], 'eq' of (27) must be unpacked into an appropriate analogue of material equivalence, and consequently the step from (28) to (29) is not valid, since the substitution would have to be made in the scope of the square. It was shown that (19) is not an unrestricted theorem in QS4.

On the other hand, since in QS4

(30) $(5+4) =_s 9$

where ' $=_s$ ' is the appropriate analogue for attributes (classes) of strict equivalence, '$5+4$' may replace '9' in (28) in accordance with (23). If, however, the square were dropped from (28), as it validly can be since

(30a) $\Box p \dashv p$

is provable, then, by the restricted (19), the very same substitution available to Sm is available here.

---

12. See my "Extensionality," *Mind*, n.s., LXIX (1960): 55–62, which overlaps to some extent the present paper.

The Interpretation of Quantification: A Substitutional View

The second prominent area of criticism of quantified modal logic involves interpretation of the operations of quantification when combined with modalities. It appears to me that at least some of the problems stem from an absence of an adequate, unequivocal, colloquial translation of the operations of quantification. It is often not quantification but our choice of reading and the implicit interpretive consequences of such a reading that lead to difficulties. Such difficulties are not confined to modal systems. The most common reading of existential quantification is

(31)   There is (exists) at least one (some) thing (person) which (who) . . .

Strawson,[13] for example, does not even admit the possibility of significant alternatives, for he says of (31): "We might think it strange that the whole of modern formal logic after it leaves the propositional logic and before it crosses the boundary into the analysis of mathematical concepts, should be confined to the elaboration of sets of rules giving the logical interrelations of formulae which, however complex, begin with these few rather strained and awkward phrases." Indeed, taking (31) at face value, Strawson gets into a muddle about tense [(30) is in the present tense], and the ambiguities of the word 'exist'. What we would like to have and do not have, is a direct, unequivocal colloquial reading of

(32) $(\exists x)\varphi x$

which gives us the force of the following:

(33)   Some substitution instance of 'φx' is true.

I am suggesting not that (33) provides translations of (32), but only that what is wanted is a translation with the force of (32).

As seen from (33), quantification has to do primarily with truth and falsity and with open sentences. Readings in accordance with (31) may entangle us unnecessarily in ontological perplexities, for if quantification has to do with things and if variables for attributes or classes can be quantified upon, then in accordance with (31) attributes and classes are things. If we still want to distinguish the identifying from the classifying function of language, then we are involved in a classification of different kinds of things and the accompanying platonic

13. *Introduction to Logical Theory.*

involvements. The solution is not to banish quantification binding variables other than individual variables but only not to be taken in by (31). We do in fact have some colloquial counterparts of (33): the nontemporal 'sometimes' or 'in some cases' or 'in at least one case', which have greater ontological neutrality than (31).

Some of the arguments involving modalities and quantification are closely connected with questions of substitution and identity. At the risk of boring my readers I will go through one of these again. In QS4 the following definitions are introduced: [14]

(34) $(\varphi =_m \psi) =_{df} (x)(\varphi x \equiv \psi x)$

(35) $(\varphi =_s \psi) =_{df} \Box(\varphi =_m \psi)$

Note that '$=_s$' and '$=_m$' represent equivalence relations weaker than identity and describable as strict and material equality. Individual descriptions can be interpreted as higher-order terms, or, if the theory of descriptions is strictly applied, the solution to some of the substitution puzzles also falls under the theorems that, as argued above, constrain substitution of material equivalents in modal contexts. Since the equality in (10) is contingent, it is the case that there is a *descriptive* reading such that

(36)   the evening star $=_m$ the morning star

It is also the case that

(37) $\Diamond \sim$(the evening star $=_m$ the morning star)

One way of writing (11) is as

(38) $\Box$(the evening star $=_m$ the evening star)

By existential generalization in (38), it follows that

(39) $(\exists \varphi)\Box(\varphi =_m$ the evening star)

In the words of (31), (39) becomes

(40)   There is a thing such that it is necessary that it is equal to the evening star.

14. See my "A Functional Calculus of First Order Based on Strict Implication" and "The Identity of Individuals in a Strict Functional Calculus." Abstracts are introduced and attributes (or classes) may be equated with abstracts. Among the obvious features of such a calculus of attributes (classes), is the presence of equivalent, nonidentical, empty attributes (classes). If the *null* attribute (class) is defined in terms of identity, it will be intersubstitutible with any abstract on a contradictory function.

The stubborn unlaid ghost rises again. Which thing, the evening star that, by (36), is equal to the morning star? But such a substitution would lead to the falsehood

(41)   □(the evening star $=_m$ the morning star)

The argument may be repeated for (27) through (29).

In QS4 the solution is clear, for, since (37) holds and since in (39) 'φ' occurs within the scope of a square, we cannot go from (39) to (41). On the other hand, the step from (38) to (39) (existential instantiation) is entirely valid, for surely there is a substituend of 'φ' for which

□($\varphi =_m$ the evening star)

is true. In particular, the case where 'φ' is replaced by 'the evening star'.

There is also the specific problem of interpreting quantification in (6) $[\Diamond (\exists x)\varphi x \; \dashv \; (\exists x)\Diamond \varphi x]$, which is a postulate of QS4. Read in accordance with (31) as

(42)   If it is logically possible that there is something that φs, then there is something such that it is logically possible that it φs.

it is admittedly odd. The antecedent seems to be about what is logically possible and the consequent about what there is. How can one go from possibility to existence? Read in accordance with (33) we have the clumsy but not so paradoxical

(43)   If it is logically possible that φx for some substituend of 'x', then there is some substituend of 'x' such that it is logically possible that φx.

Although the emphasis has now been shifted from things to statements, and the ontological consequences of (42) are absent, (43) is still indirect and awkward. It would appear that questions such as the acceptability or nonacceptability of (6) are best solved in terms of some semantical construction. We will return to this, but first some minor matters.

Modalities Misunderstood

A defense of modal logic would be incomplete without touching on criticisms of modalities that stem from confusion about what is or is not provable in such systems. One criticism is that of Paul Rosenbloom,[15] who seized on my proof[16] that a strong deduction theorem is not available in QS4 as a reason for discarding strict implication as relevant in any way to the deducibility relation. Rosenbloom failed to note that a weaker and perhaps more appropriate deduction theorem is available. Indeed, Anderson and Belnap,[17] in their attempt to formalize entailment without modalities, reject the strong form of the deduction theorem as "counter-intuitive for entailment."

Another example occurs in *Word and Object*;[18] it can be summarized as follows:

(44) Modalities yield talk of a difference between necessary and contingent attributes.

(45) Mathematicians may be said to be necessarily rational and not necessarily two-legged.

(46) Cyclists are necessarily two-legged and not necessarily rational.

(47) *a* is a mathematician and a cyclist.

(48) Is this concrete individual necessarily rational and contingently two-legged or vice versa?

(49) "Talking referentially of the object, with no special bias toward a background grouping of mathematicians as against cyclists . . . there is no semblance of sense in rating some of his attributes as necessary and others as contingent."

Quine says that (44) to (47) are supposed to "evoke the appropriate sense of bewilderment," and they surely do, for I know of no interpreted modal system that countenances necessary attributes in the manner suggested. Translating (45) to (47), we have one of the equivalent

(50) $(x)(Mx \prec Rx) \equiv (x)\Box(Mx \supset Rx) \equiv (x) \sim \Diamond (Mx \cdot \sim Rx)$

conjoined in (45) with one of the equivalent

(51) $(x) \sim \Box(Mx \supset Tx)$

15. *The Elements of Mathematical Logic* (New York: Dover, 1950), p. 60.
16. "Strict Implication, Deducibility, and the Deduction Theorem," *Journal of Symbolic Logic,* XVIII (1953): 234–236.
17. A. R. Anderson and N. D. Belnap, *The Pure Calculus of Entailment* (pre-print).
18. *Word and Object,* pp. 199–200.

$$\equiv (x) \diamond \sim (Mx \supset Tx)$$
$$\equiv (x) \diamond (Mx \cdot \sim Tx)$$

Also one of the equivalent

(52) $(x)(Cx \dashv 3 \ Tx) \equiv (x) \ \Box \ (Cx \supset Tx) \equiv (x) \sim \diamond (Cx \cdot \sim Tx)$

conjoined in (46) with one of the equivalent

(53) $(x) \sim \Box (Cx \supset Rx) \equiv (x) \diamond \sim (Cx \supset Rx)$
$$\equiv (x) \diamond (Cx \cdot \sim Rx)$$

And in (47)

(54) $Ma \cdot Ca$

Among the conclusions we can draw from (49) to (54) are $\Box(Ma \supset Ra)$, $\sim \diamond (Ma \cdot \sim Ra)$, $\diamond (Ma \cdot \sim Ta)$, $\sim \Box (Ma \supset Ta)$, $\Box(Ca \supset Ta)$, $\sim \diamond (Ca \cdot \sim Ta)$, $\diamond (Ca \cdot \sim Ra)$, $\sim \Box (Ca \cdot \sim Ra)$, Ta, Ra, $Ta \cdot Ra$; but nothing to answer question (48), or to make any sense of (49). Quine would appear to be assuming that

(55) $(p \dashv 3 \ q) \dashv 3 \ (p \dashv 3 \ \Box \ q)$

is provable in QS4, but it is not so provable, except where $p \equiv \Box r$ for some r. Keeping in mind that, if we are dealing with logical modalities, we can see that none of the attributes (M, R, T, C) in (50) to (54), taken separately or conjoined, are logically necessary. These are not the sort of attributes that modal logic, even derivatively, countenances as being logically necessary. But a word is appropriate here about the derivative sense in which we *can* speak of necessary and contingent attributes.

In QS4 abstracts are introduced such that to every function there corresponds an abstract, e.g.,

(56) $x \varepsilon \hat{y} A =_{df} B$

where B is the result of substituting every free occurrence of y in A by x, given familiar constraints.

If r is some abstract, we can define

(57) $x \varepsilon \ \boxdot \ r =_{df} \Box(x \varepsilon r)$ $\vdash \boxdot \ r =_{df} (x)(x \varepsilon \boxdot r)$

and

(58) $x \varepsilon \ \diamond \ r =_{df} \diamond (x \varepsilon r)$ $\vdash \diamond \ r =_{df} (x)(x \in \diamond r)$

It is clear that among the abstracts to which $\vdash \boxdot$ may validly be affixed will be those corresponding to valid functions, e.g., $\hat{y}(y I y)$, $\hat{y}(\varphi x$

$\vee \sim \phi x$), etc. It would be appropriate to call these *necessary* attributes, and the symbol '$\square$' is a derivative way of applying modalities to attributes.

Similarly, all the constituent attributes of (50) to (54) could in the sense of (58) be called possible, where '$\diamond$' is the derivative modality for possibility of attributes. A contingent attribute would be possible but not necessary. However, *if* (50) is true, then the attribute of being either not a mathematician or rational could appropriately be called necessary, for it would follow from the premises that

(59) $(x)\square(x\varepsilon\hat{y}(\sim My \vee Ry))$

but some ground for importing the necessity of the premise would be required.

Semantic Constructions

I would like in conclusion to suggest that the polemics of modal logic are perhaps best carried out in terms of some explicit semantical construction. As we know from disputes about interpretation in connection with (6) $[\diamond(\exists x)A \dashv 3 (\exists x) \diamond A]$, to argue without some construction is awkward at best and at worst has the character of a quibble.

Let us reappraise (6) in terms of such a construction.[19] Consider, for example, a language (*L*), with truth-functional connectives, a modal operator ($\diamond$), a finite number of individual constants, an infinite number of individual variables, one two-place predicate (*R*), quantification, and the usual criteria for being well-formed. A domain (*D*) of individuals is then considered, named by the constants of *L*. A model of *L* is defined as a class of ordered couples (possibly empty) of *D*. The members of a model of *L* are exactly those pairs between which *R* holds. To say, therefore, that the atomic sentence $R(a_1 a_2)$ of *L* holds or is true in *M* is to say that the ordered couple (b_1, b_2) is a member of *M*, where a_1 and a_2 are the names in *L* of b_1 and b_2. If a sentence *A* of *L* is of the form $\sim B$, *A* is true in *M* if and only if *B* is not true in *M*. If *A* is of the form $(B_1 \cdot B_2)$, then *A* is true in *M* if and only if both B_1 and B_2 are true in *M*. If *A* is of the form $(\exists x)B$, then *A* is true in *M* if and only if at least one substitution in-

---

19. The construction here outlined is close to that of R. Carnap, *Meaning and Necessity* (Chicago: University of Chicago Press, 1947). The statement of the construction is in accordance with a method of J. C. C. McKinsey. See also McKinsey, "On the Syntactical Construction of Systems of Modal Logic," *Journal of Symbolic Logic*, X (1946): 88–94; "A New Definition of Truth," *Synthese*, VII (1948/49): 428–433.

stance of B is true (holds) in M. If A is $\Diamond\, B$ then A is true in M if and only if B is true in some model M_1.

We see on the construction that a true sentence of L is defined relative to a model and a domain of individuals. A logically true sentence is a sentence that would be true in every model. We are now in a position to give a rough proof of (6). Suppose (6) is false in some M. Then

$$\sim\Diamond\,(\,\Diamond\,(\exists x)\varphi x\cdot\sim(\exists x)\,\Diamond\,\varphi x)$$

is false in M. Therefore

$$\Diamond\,(\,\Diamond\,(\exists x)\varphi x\cdot\sim(\exists x)\,\Diamond\,\varphi x)$$

is true in M. So

$$\Diamond\,(\exists x)\phi x\cdot\sim(\exists x)\,\Diamond\,\phi x$$

is true in some M_1. Therefore

(60) $\Diamond\,(\exists x)\varphi x$

and

(61) $\sim(\exists x)\,\Diamond\,\varphi x$

are true in M_1. Consequently, from (60),

(62) $(\exists x)\varphi x$

is true in some model M_2. Therefore there is a member of D (b) such that

(63) φb

is true in M_2. But from (61)

$$(\exists x)\,\Diamond\,\varphi x$$

is not true in M_1. Consequently, there is no member b of D such that

(64) $\Diamond\,\varphi b$

is true in M_1. So there is no model M_2 such that

$$\varphi b$$

is true in M_2. But this result contradicts (63). Consequently, in such a construction, (6) must be true in every model.

If this is the sort of construction one has in mind, then we are persuaded of the plausibility of (6). Indeed, going back to (43), it can be seen that this was the sort of construction that was being assumed.

If (6) is to be regarded as offensive, it must be in terms of some other semantic construction that ought to be made explicit.[20]

We see, that, though the rough outline above corresponds to the Leibnizian distinction between true in a possible world and true in all possible worlds, it is also to be noted that there are no specifically intensional objects. No new entity is spawned in a possible world that isn't already in the domain in terms of which the class of models is defined. In such a model modal operators have to do with truth relative to the model. On this interpretation,[21] Quine's "flight from intension" may have been exhilarating, but unnecessary.

20. A criticism of the construction here outlined is the assumption of the countability of members of *D*. McKinsey points this out in the one chapter I have seen (chap. 1, vol. 2) of a projected (unpublished) two-volume study of modal logic, and indicates that his construction will not assume the countability of members of *D*. Whereas Carnap's construction leads to a system at least as strong as S5, McKinsey's (he claims) will be at least as strong as S4. I have not seen or been able to locate any parts of this study, in which the details were to have been worked out along with completeness proofs for some of the Lewis systems. See also J. Myhill, in *Logique et analyse* (1958), pp. 74–83; and S. Kripke, *Journal of Symbolic Logic,* XXIV (1959): 323–324 (abstract).

21. If one wishes to talk about possible things, then of course such a construction is inadequate.

APPENDIX 1A: DISCUSSION

This appendix appeared in *Synthese*, XIV (September 1962). It includes comments of some of the participants in the Boston Colloquium for the Philosophy of Science on the occasion of the presentation of "Modalities and Intensional Languages" in February 1962. The transcript of the tape was circulated to the participants for final editing. Not all comments of participants were included in the published version.

The present printing includes some editorial corrections and a footnote to a passage that needs clarification and correction.

I debated as to whether the discussion should be included in the present volume but decided that it had sufficient interest, including historical interest, to be worth inclusion. ■

PROF. MARCUS: We seem still at the impasse I thought to resolve at this time. The argument concerning (12) was informal, and parallels as I suggested, questions raised in connection with the 'paradox' of analysis. One would expect that if a statement were analytic, and it bore a strong equivalence relation to a second statement, the latter would be analytic as well. Since (12) cannot be represented in Sm without restriction, the argument reveals material equivalence to be insufficient and weak. An adequate representation of (12) requires a modal framework.

The question I have about essentialism is this: Suppose these modal systems are extended in the manner of *Principia* to higher order. Then

$$\Box((5+4)=9)$$

will hold. Here ' $=$ ' may be taken as either ' $=_s$' or ' $=_m$' of the present paper, (since the reiterated squares telescope), whereas

$$\Box((5+4)=\text{the number of planets})$$

does not hold. Our interpretation of these results commits us only to the conclusion that the equivalence relation that holds between $5+4$

and 9 is stronger than the one that holds between $5 + 4$ and the number of planets. More specifically, the stronger one is the class or attribute analogue of $\equiv$. No mysterious property is being conferred on either 9 or the number of planets that they do not already have in the extensional $((5 + 4) =_m$ the number of planets).

PROF. QUINE: May I ask if Kripke has an answer to this? . . . Or I'll answer, or try to.

MR. KRIPKE: As I understand Professor Quine's essentialism, it isn't what's involved in either of these two things you wrote on the board that causes trouble. It is in inferring that there exists an x, which necessarily $= 5 + 4$ (from the first of the two). (To Quine:) Isn't that what's at issue?

PROF. QUINE: Yes.

MR. KRIPKE: So this attributes necessarily equalling $5 + 4$ to an object.

PROF. MARCUS: But that depends on the suggested interpretation of quantification. We prefer a reading that is not in accordance with things, unless, as in the first-order language, there are other reasons for reading in accordance with things.

PROF. QUINE: That's true.

PROF. MARCUS: So the question of essentialism arises only on your reading of quantification. For you, the notion of reference is univocal, absolute, and bound up with the expressions, of whatever level, on which quantification is allowed. What I am suggesting is a point of view that is not new to the history of philosophy and logic: That all terms may "refer" to objects, but that not all objects are things, where a thing is at least that about which it is appropriate to assert the identity relation. We note a certain historical consistency here, as, for example, the reluctance to allow identity as a relation proper to propositions. If one wishes, one could say that object-reference (in terms of quantification) is a wider notion than thing-reference, the latter being also bound up with identity and perhaps with other restrictions as well, such as spatiotemporal location. If one wishes to use the word 'refer' exclusively for thing-reference, then we would distinguish those names that refer, from those that name other sorts of objects. Considered in terms of the semantical construction proposed at the end of the paper, identity is a relation that holds between individuals; and their names have thing-reference. To say of a thing a that it necessarily has a property φ, $(\Box(\varphi a))$, is to say that φa is true in every model. Self-identity would be such a property.

PROF. QUINE: Speaking of the objects or the referential end of things in terms of identity, rather than quantification, is agreeable to me in the sense that for me these are interdefinable anyway. But what's appropriately regarded as the identity matrix, or open sentence, in the theory is for me determined certainly by consideration of quantification. Quantification is a little bit broader, a little bit more generally applicable to the theory because you don't always have anything that would fulfill this identity requirement. As to where essentialism comes in: what I have in mind is an interpretation of this quantification where you have an x here (in $\Box((5+4)=x)$). Now, I appreciate that from the point of view of modal logic, and of things that have been done in modal logic in Professor Marcus's pioneer system, this would be regarded as true rather than false:

$$\Box((5+4)=9)$$

This is my point, in spite of the fact that if you think of this ($\Box((5+4)=$ the number of planets)) as what it is generalized from, it ought to be false.

PROF. MARCUS: $\Box((5+4)=$ the number of planets) *would* be false. But this does not preclude the truth of

$$(\exists x)\Box((5+4)=x)$$

any more than the falsehood

$$13 = \text{the number of Christ's disciples}$$

precludes the truth of

$$(\exists x)(13=x)$$

(We would, of course, take ' $=$ ' as ' $=_m$ ' here.)

PROF. QUINE: That's if we use quantification in the ordinary ontological way and that's why I say we put a premium on the *nine* as over against the *number of planets*; we say this term is what is going to be *maßgebend* for the truth value of this sentence in spite of the fact that we get the opposite whenever we consider the other term. This is the sort of specification of the number that counts:

$$5+4=9$$

This is not:

$$5+4=\text{number of planets}$$

I grant further that essentialism does not come in if we interpret quantification in your new way. By quantification I mean quantification in the ordinary sense rather than a new interpretation that might fit most if not all of the formal laws that the old quantification fits. I say 'if not all', because I think of the example of real numbers again. If on the other hand we do not have quantification in the old sense, then I have nothing to suggest at this point about the ontological implications or difficulties of modal logic. The question of ontology wouldn't arise if there were no quantification of the ordinary sort. Furthermore, essentialism certainly wouldn't be to the point, for the essentialism I'm talking about is essentialism in the sense that talks about objects, certain objects; that an object has certain of these attributes essentially, certain others only accidentally. And no such question of essentialism arises if we are only talking of the terms and not the objects that they allegedly refer to. Now, Professor Marcus also suggested that possibly the interpretation could be made something of a hybrid between the two—between quantification thought of as a formal matter, and just talking in a manner whose truth conditions are set up in terms of the expression substituted rather than in terms of the objects talked about; and that there are other cases where we can still give quantification the same old force. Now, that may well be: we might find that in the ordinary sense of quantification I've been talking about there is quantification into nonmodal contexts and no quantification but only this sort of quasi-quantification into the modal ones. And this conceivably might be as good a way of handling such modal matters as any.

PROF. MARCUS: It is not merely a way of coping with perplexities associated with intensional contexts. I think of it as a better way of handling quantification.

You've raised a problem that has to do with the real numbers. Perhaps the Cantorian assumption is one we can abandon. We need not be particularly concerned with it here.

PROF. QUINE: It's one thing I would certainly be glad to avoid, if we can get all of the classical mathematics that we do want.

MR. KRIPKE: This is what I thought the issue conceivably might be, and hence I'll raise it explicitly in this form: Suppose this system contains names, and suppose the variables are supposed to range over numbers, and using '9' as the name of the number of planets, and the usual stock of numerals, '0', '1', '2' . . . and in addition various other primitive terms for numbers, one of which would be 'NP' for the 'number of planets', and suppose '$\Box(9>7)$' is true, according to our system. But say we also have '$\sim\Box(NP>7)$'. Now suppose 'NP' is taken to be as legitimate a name for the number of planets as

'9' (i.e., for this *number*) as the numeral itself. Then we get the odd-seeming conclusion (anyway in your [Marcus's] quantification) that

$$(\exists x, y)(x = y \cdot \Box(x > 7) \cdot \sim \Box(y > 7))$$

On the other hand, if 'NP' is not taken to be as legitimate a name for the number of planets as '9', then, in that case, I presume that Quine would reply that this sort of distinction amounts to the distinction of essentialism itself. (To Quine:) Would this be a good way of stating your position?

PROF. QUINE: Yes. And I think this formula is one that Professor Marcus would accept under a new version of quantification. Is that right?

PROF. MARCUS: No . . . this wouldn't be true under my interpretation, if the ' = ' (of Kripke's expression) is taken as identity. If it were taken as identity, it would be not only odd-seeming but contradictory. If it is taken as ' $=_m$ ', then it is not odd-seeming but true. What we must be clear about is that in the extended modal systems with which we are dealing here, we are working within the framework of the theory of types. On the level of individuals, we have only identity as an equivalence relation between individuals. On the level of predicates, or attributes, or classes, or propositions, there are other equivalence relations that are weaker. Now, the misleading aspect of your [Kripke's] formulation is that when you say, "Let the variables range over the numbers," we seem to be talking about individual variables, ' = ' must then name the identity relation and we are in a quandary. But within a type framework, if *x* and *y* can be replaced by names of numbers, then they are higher-type variables and the weaker equivalence relations are appropriate in such contexts.

MR. KRIPKE: Well, you're presupposing something like the Frege-Russell definition of number, then?

PROF. MARCUS: All right. Suppose numbers are generated as in *Principia* and suppose 'the number of planets' may be properly equated with '9'. The precise nature of this equivalence will of course depend on whether 'the number of planets' is interpreted as a description or a predicate, but in any case, it will be a much weaker equivalence.

MR. KRIPKE: Nine and the number of planets do not in fact turn out to be identically the same?

PROF. MARCUS: No, they're not. That's just the point.

MR. KRIPKE: Now, do you admit the notion of "identically the same" at all?

PROF. MARCUS: That's a different question. I admit identity on the level of individuals certainly. Nor do I foresee any difficulty in allowing the identity relation to hold for objects named by higher-type expressions (except perhaps propositional expressions), other than the ontological consequences discussed in the paper. What I am *not* admitting is that "identically the same" is indistinguishable from weaker forms of equivalence. It is explicit or implicit extensionalizing principles that obliterate the distinction. On this analysis, we could assert that

9 is identically the same as 9

but not

9 is identically the same as (5 + 4)

without some very weak extensionalizing principle that reduces identity to logical equivalence.

MR. KRIPKE: Supposing you have any identity, and you have something varying over individuals.

PROF. MARCUS: In the theory of types, numbers are values for predicate variables of a kind to which several equivalence relations are proper.

MR. KRIPKE: Then, in your opinion the use of *numbers* (rather than individuals) in my example is very important.

PROF. MARCUS: It's crucial.

PROF. QUINE: That's what I used to think before I discovered the error in Church's criticism. And if I understand you, you're suggesting now what I used to think was necessary; namely, in order to set these things up, we're going to have, as the values of variables, not numbers, but assorted number properties, that are equal, but different numbers—the number of planets on the one hand, 9 on the other. What I say now is that this proliferation of entities isn't going to work. For example, take x as just as narrow and intensional an object as you like . . .

PROF. MARCUS: Yes, but not on the level of individuals, where only one equivalence relation is present. (We are omitting here consideration of such relations as congruence.)

PROF. QUINE: No, my x isn't an individual. The values of 'x' may be properties, or attributes, or propositions, that is, as intensional as you like. I argue that if $\varphi(x)$ determines x uniquely, and if p is not implied by $\varphi(x)$, still the conjunction $p \cdot \varphi(x)$ will determine that same highly abstract attribute, or whatever it was, uniquely, and yet these two conditions will not be equivalent, and therefore this kind of argument can be repeated for it. My point is, we can't get out of the difficulty by splitting up the entities; we're going to have to get out of it by essentialism. I think essentialism, from the point of view of the modal logician, is something that ought to be welcome. I don't take this as being a *reductio ad absurdum.*

PROF. MCCARTHY: (MIT): It seems to me you can't get out of the difficulty by making 9 come out to be a class. Even if you admit your individuals to be much more inclusive than numbers. For example, if you let them be truth values. Suppose you take the truth value of the 'number of planets is nine', then this is something which is true, which has the value truth. But you would be in exactly the same situation here. If you carry out the same problem, you will still get something that will be 'there exists x, y such that $x = y$ and it is necessary that x is true, but it is not necessary that y is true'.

PROF. MARCUS: In the type framework, the individuals are neither numbers, nor truth values, nor any object named by higher-type expressions. Nor are the values of sentential variables truth values. Sentential or propositional variables take as substituends sentences (statements, names of propositions if you will). As for your example, there is no paradox, since your ' = ' would be a material equivalence, and by virtue of the substitution theorem, we could not replace 'x' by 'y' in '$\Box \, x$' (y being contingently true).

PROF. MCCARTHY: Then you don't have to split up numbers, regarding them as predicates either, unless you also regard truth functions as predicates.

PROF. MARCUS: About ''splitting up''. If we must talk about objects, then we could say that the objects in the domain of individuals are extensions, and the objects named by higher-order expressions are intensions. If one is going to classify objects in terms of the intension—extension dualism, then this is the better way of doing it. It appears to me that a failing of the Carnap approach to such questions, and one that generated some of these difficulties, is the passion for symmetry. Every term (or name) must, according to Carnap, have a dual role. To me it seems unnecessary and does proliferate entities unnecessarily. The kind of evidence relevant here is informal. We do, for example, have a certain hesitation about talking of identity of propositions, and we do acknowledge a certain difference between talking of identity of attributes as against identity in connection with individuals. And to speak of the intension named by a proper name strikes one immediately as a distortion for the sake of symmetry.

FØLLESDAL: The main question I have to ask relates to your argument against Quine's examples about mathematicians and cyclists.

You say that (55) is not provable in QS4. Is your answer to Quine that it is not provable?

PROF. MARCUS: No. My answer to Quine is that I know of no modal system, extended of course, to include the truth of:

It is necessary that mathematicians are rational.

and

It is necessary that cyclists are two-legged.

by virtue of meaning postulates or some such, where his argument applies. Surely if the argument was intended as a criticism of modal logic, as it seems to be, he must have had *some* formalization in mind, in which such paradoxes might arise.

FØLLESDAL: It seems to me that the question is not whether the formula is provable, but whether it's a well-formed formula, and whether it's meaningful.

PROF. MARCUS: The formula in question is entirely meaningful, well formed if you like, given appropriate meaning postulates (defining statements) that entail the necessity of

All mathematicians are rational.

and

All cyclists are two-legged.

I merely indicated that there would be no way of *deriving* from these meaning postulates (or defining statements) as embedded in a modal logic, anything like

It is necessary that John is rational.

given the truth:

John is a mathematician.

although both statements are well formed and the relation between 'mathematician' and 'rational' is analytic. The paradox simply does not arise. What I *did* say is that there is a derivative sense in which one can talk about necessary attributes, in the way that abstraction is derivative. For example, since it is true that

$(x)\Box(xLx)$

which with abstraction gives us

$$(x)\square(x \in \hat{y}(yIy))$$

which as we said before, would give us

$$\vdash\square]\hat{y}(yIy))$$

The property of self-identity may be said to be necessary, for it corresponds to a tautological function. Returning now to Professor Quine's example, if we introduced constants like 'cyclist', 'mathematician', etc., and appropriate meaning postulates, then the attributes of being either a nonmathematician or rational would also be necessary. Necessary attributes would correspond to analytic functions in the broader sense of analytic. These may be thought of as a kind of essential attribute, although necessary attribute is better here. For these are attributes that belong necessarily to every object in the domain, whereas the usual meaning of essentialism is more restricted. Attributes like mathematician and cylist do not correspond to analytic functions.

PROF. QUINE: I've never said or, I'm sure, written that essentialism could be proved in any system of modal logic whatever. I've never even meant to suggest that any modal logician even was aware of the essentialism he was committing himself to, even implicitly in the sense of putting it into his axioms. I'm talking about quite another thing— I'm not talking about theorems, I'm talking about truth, I'm talking about true interpretation. And what I have been arguing is that if one is to quantify into modal contexts and one is to interpret these modal contexts in the ordinary modal way and one is to interpret quantification as quantification, not in some quasi-quantificatory way that puts the truth conditions in terms of substitutions of expressions, then in order to get a coherent interpretation one has to adopt essentialism and I already explained a while ago just how that comes about. But I did not say that it could ever be deduced in any of the S-systems or any system I've ever seen.

PROF. MARCUS: I was not suggesting that you contended that essentialism could be proved in any system of modal logic. But only that I know of no interpreted modal system, even where extended to include predicate constants such as those of your examples, where properties like being a mathematician would necessarily belong to an object. The kind of uses to which *logical* modalities are put have nothing to do with essential properties in the old ontological sense. The introduction of physical modalities would bring us closer to this sort of essentialism.

32

FØLLESDAL: That's what creates the trouble when one thinks about properties of this kind, like being a cyclist.

PROF. QUINE: But then you can't use quantifiers as quantifiers.

PROF. MARCUS: The interpretation of quantification has advantages other than those in connection with modalities. For example, many of the perplexities in connection with quantification raised by Strawson in *Introduction to Logical Theory* are clarified by the proposed reading of quantification. Nor is it my conception. One has only to turn to the introduction of *Principia Mathematica* where universal and existential quantification is discussed in terms of 'always true' and 'sometimes true'. It is a way of looking at quantification that has been neglected. Its neglect is a consequence of the absence of a uniform, colloquial way of translating, although we can always find some adequate locution in different classes of cases. It is *easier* to say, ''There is a thing which . . .'' and since it is adequate some of the time it has come to be used universally with unfortunate consequences.

PROF. QUINE: Well, Frege, who started quantification theory, had the regular ontological interpretation. Whitehead and Russell fouled it up because they confused use and mention.

FØLLESDAL:: It seems from the semantical considerations that you have at the end of the paper, that you need your special axiom.

PROF. MARCUS: Yes, for that construction. I have no strong preferences. It would depend on the uses to which some particular modal system is to be put.

FØLLESDAL:: You think you might have other constructions?

PROF. MARCUS: Indeed. Kripke, for example, has suggested other constructions. My use of this particular construction is to suggest that in discussions of the kind we are having here today, and in connection with the type of criticism raised by Professor Quine in *Word and Object* and elsewhere, it is perhaps best carried out with respect to some construction.

MR. KRIPKE: Forgetting the example of numbers, and using your interpretation of quantification—there's nothing seriously wrong with it at all—does it not require that for any two names, '*A*' and '*B*', of individuals, '*A* = *B*' should be *necessary,* if true at all? And if '*A*' and '*B*' are names of the same individual, that any necessary statement containing 'A' should remain necessary if '*A*' is replaced by '*B*'?

PROF. MARCUS: We might want to say that for the sake of clarity and ease of communication, it would be convenient if to each object there were attached a single name. But we can and we do attach more than one name to a single object. We are here talking of proper names in the ideal sense, as tags and not descriptions. Presumably, if a single object had more than one tag, there would be a way of finding out, such as having recourse to a dictionary[1] or some analogous inquiry, which would resolve the question as to whether the two tags denote the same thing. If 'Evening Star' and 'Morning Star' are considered to be two proper names for Venus, then finding out that they name the same thing that 'Venus' names is different from finding out what is Venus's mass, or its orbit. It is perhaps admirably flexible, but also very confusing, to obliterate the distinction between such linguistic and properly empirical procedures.

MR. KRIPKE: That seems to me like a perfectly valid point of view. It seems to me the only thing Professor Quine would be able to say and therefore what he must say, I hope, is that the assumption of a distinction between tags and empirical descriptions, such that the truth-values of identity statements between tags (but not between descriptions) are ascertainable merely by recourse to a dictionary, amounts to essentialism itself. The tags are the "essential" denoting phrases for individuals, but empirical descriptions are not, and thus we look to statements containing "tags", not descriptions, to ascertain the essential properties of individuals. Thus the assumption of a distinction between "names" and "descriptions" is equivalent to essentialism.

PROF. QUINE: My answer is that this kind of consideration is not revelant to the problem of essentialism because one doesn't ever need descriptions or proper names. If you have notations consisting of simply propositional functions (that is to say, predicates) and quantifiable variables and truth functions, the whole problem remains. The distinction between proper names and descriptions is a red herring. So are the tags. (Marcus: Oh, no.)

All it is is a question of open sentences that uniquely determine. We can get this trouble every time, as I proved with my completely general argument of p in conjunction with φx where x can be as finely discriminated an intension as one pleases—and in this there's no singular term at all except the quantifiable variables or pronouns themselves. This was my answer to Smullyan years ago, and it seems to me the answer now.

MR. KRIPKE: Yes, but you have to allow the writer what she herself says, you see, rather than arguing from the point of view of your own interpretation of the quantifiers.

---

1. Since such entries are usually described as "nonlexical", the dictionary here functions as an encyclopedia. But this whole passage needs clarification and emendation. As indicated in the earlier text, discovering that Hesperus and Phosphorus have the same path is an empirical discovery that entails that they are identical and hence that 'Hesperus' and 'Phosphorus' name the same thing. But the identity, once given, is necessary [added 1991].

PROF. QUINE: But that changes the subject, doesn't it? I think there are many ways you can interpret modal logic. I think it's been done. Prior has tried it in terms of time and one thing and another. I think any consistent system can be found an intelligible interpretation. What I've been talking about is quantifying, in the quantificational sense of quantification, into modal contexts in a modal sense of modality.

MR. KRIPKE: Suppose the assumption in question is right—that every object is associated with a tag, which is either unique or unique up to the fact that substituting one for the other does not change necessities—is that correct? Now, then granted this, why not read "there exists an *x* such that necessarily *p* of *x*" as (put in an ontological way if you like) "there exists an object x with a name *a* such that *p* of *a* is analytic." Once we have this notion of name, it seems unexceptionable.

PROF. QUINE: It's not very far from the thing I was urging about certain ways of specifying these objects being by essential attributes and that's the role that you're making your attributes play.

MR. KRIPKE: So, as I was saying, such an assumption of names is equivalent to essentialism.

PROF. COHEN: I think this is a good friendly note on which to stop.

APPENDIX 1B: SMULLYAN ON MODALITY AND DESCRIPTION

The following commentary is taken from a review in the *Journal of Symbolic Logic,* XIII (1948): 149–150. ■

In his paper "Modality and Description" (*Journal of Symbolic Logic,* XIII (1948): 31–37) Arthur F. Smullyan is concerned with an antinomy that allegedly arises from substituting equals for equals in modal contexts. Quine for example has argued that the true premises

A.   It is logically necessary that 9 is less than 10.

B.   9 = the number of planets.

lead to the false conclusion

C.   It is logically necessary that the number of planets is less than 10.

Smullyan attempts to resolve this dilemma within a logical system such as that of *Principia Mathematica* that has been appropriately extended to include modal operators. He presents two analogous solutions. In the first, the expression

D.   the number of planets

abbreviates a descriptive phrase, and descriptions are treated in accordance with *14 of *Principia Mathematica*. In the second, D is interpreted as an abstract, and abstracts are introduced in a manner similar to that of Russell.

Where D is construed as a description, A, B, C can be regarded as an instance of

A'. $N(Fy)$

B'. $y = (\iota x)(\phi x)$

C'. $N[F(\iota x)(\phi x)]$

Smullyan shows that at most, A' and B' yield

$$(\exists x)((\phi z) \equiv_z z = x \cdot N(Fx))$$

which is equivalent to

$$[(\imath x)(\phi x)] \cdot N(F(\imath x)(\phi x))$$

If D is interpreted as an abstract, the method of solution is similar, provided that the definition employed is unambiguous with respect to the scope of the abstract. Smullyan chooses the definition

E. $[\hat{x}(Gx)] \cdot F\hat{x}(G\hat{x}) =_{df} (\exists \alpha)(Gx \equiv_x x \in \alpha \cdot Fa)$
    where α is a class variable

A, B, C may then be regarded as illustrating

A″. $N[f\hat{x}(Ax)]$

B″. $\hat{x}(Ax) = \hat{x}(Bx)$

C″. $N[f\hat{x}(Bx)]$

If A″, B″, and C″ are expanded in accordance with E, it is apparent that C″ cannot be inferred from the premises A″ and B″.

In the reviewer's opinion, Smullyan is justified in his contention that the solution of Quine's dilemma does not require any radical departure from a system such as that of *Principia Mathematica*. Indeed, since such a solution is available, it would seem to be an argument in favor of Russell's method of introducing abstracts and descriptions.

If modal systems are not to be entirely rejected, a solution of the dilemma A, B, C requires that the equality relation that holds between expressions such as '9' and 'the number of planets' must be distinguished from the equality relation that holds, for example, between the expressions '9' and '7+2'. This distinction must be such that '7+2' may replace '9' in modal contexts but 'the number of planets' may not. In a system of the kind with which Smullyan is presumably concerned, an unrestricted substitution theorem can be proved only for expressions that are necessarily equivalent. (See the reviewer's paper, *Journal of Symbolic Logic*, XI [1946]). Thus where E is construed in terms of abstracts (and this is the more natural interpretation), the deduction of C″ from A″ and B″ is precluded by the aforementioned restriction on substitution. A″ is an abbreviation for

$$N(\exists \alpha)(Ax \equiv_x x \in \alpha \cdot f\alpha)$$

and B″ is equivalent to

$$Ax \equiv_x Bx$$

Since 'Ax' comes within the scope of a modal operator in A″, 'Ax' cannot be replaced by 'Bx'.

Although Smullyan constructs his modal system informally, his assumption

S2. $N[x \in \alpha \equiv_x x \in \beta \supset \alpha = \beta]$

(p. 36) seems to lead to the result that if two classes have the same members, then it is necessary that they have the same members. Let '$A \dashv B$' and '$A \equiv B$' abbreviate '$N(A \supset B)$' and '$N(A \equiv B)$' respectively. The major steps in the proof of this result are as follows: $N(\alpha = \alpha)$, $(\alpha = \beta) \dashv (N(\alpha = \alpha) \supset N(\alpha = \beta))$, $(\alpha = \beta) \dashv N(\alpha = \beta)$, $(x \in \alpha \equiv_x x \in \beta) \equiv (\alpha = \beta)$, $(x \in \alpha \equiv_x x \in \beta) \dashv N(x \in \alpha \equiv_x x \in \beta)$.

S2 could be replaced by

S2′. $N(x \in \alpha \equiv_x x \in \beta) \supset (\alpha = \beta)$

without altering Smullyan's general analysis. If S2′ were assumed in place of S2, it would not be necessary to reject

S3. $(\exists \alpha)N(\phi x \equiv_x x \in \alpha)$

as an assumption that leads to seemingly paradoxical results.

2. *Iterated Deontic Modalities*

The source of this paper appeared in *Mind,* LXXV, n.s. 300 (1966): 580–582. Although deontic logic has had a considerable evolution and refinement since that time, many of the problems of interpretation remain. ■

In several of the formalized systems of deontic logic, the class of well-formed formulas includes sentences in which deontic modal operators occur within the scope of others. Such repetitions are not generally eliminable by means of strong substitution rules. Some iterations, such as in 'It is obligatory that it is obligatory that everyone keep his promises', seem a kind of stuttering, and it might be argued that deontic logic should provide for their exclusion. Yet others like

(1)   Parking on highways ought to be forbidden.

make good sense. Furthermore, it is claimed that there are some theses involving nested occurrences of the deontic operators that are truths of deontic logic. One of them

(2) $O(OA \supset A)$

may be read as 'It ought to be the case that what ought to be the case is the case'. A. N. Prior[1] claims that (2) is intuitively acceptable.

The problem of interpreting multiple occurrences of the deontic operators is significant because it reveals that at least two uses of 'ought' are confounded in the arguments that are supposed to give intuitive grounds for theses such as (2). Consider the colloquial reading of the parenthetic clause of (2): 'What ought to be the case is the case'. This clause is descriptive of a state of affairs, and, contrary to Leibniz, we take it to be false. But, true or false, it does not describe an action that can be prescribed or enjoined. When we assert that it ought to be the kind of world where a person's actions always flow from his obligations, this expresses our belief about a state of affairs that, if it obtained, would make for a better world. Such beliefs may provide reasons for justifying a prescription, but they are not themselves prescriptions.

The ambiguity of interpretation is made apparent if we recall that 'obligatory' (O), 'forbidden' (F), 'permitted' (P) are taken to be interdefinable, and fall into a square of opposition. Furthermore, 'It is obligatory that' and 'It ought to be the case that' are taken to be equivalent alternative readings, for consider an instance of (2) that is supposed to be a deontic truth: 'It ought to be the case that if everyone ought to be kind then everyone is kind' and that is held to be equivalent to 'It is forbidden that everyone ought to be kind and not everyone is kind'. The latter is either peculiar or false. How can it be *forbidden* that such a state of affairs prevail unless the interdiction is part of a blueprint for creation?

1. A. N. Prior, *Formal Logic* (New York: Oxford University Press, 1962), p. 225.

We do not use 'forbidden' in the more general evaluative sense to reflect our beliefs about what is or is not desirable. 'Forbidden' is more clearly prescriptive of action, and 'obligatory' its more proper contrary. Attending to this distinction, we can appraise the previous example (1) of intelligible iteration, which in terms of the deontic modalities may be written as

(3)   O(FA)

taking 'A' as 'People park on highways'. Implicit in (1) is the belief that the consequence of prohibiting parking on highways will lead to a desirable state of affairs. Yet consider the supposed equivalent of (1):

(4)   It is forbidden that parking is (be) permitted on highways.

or, in terms of the deontic modalities,

(5)   F(PA)

An apparent paradox, for (4) seems to describe the very state of affairs deemed desirable, as if it already obtained. We may conclude that the 'ought' of (1) and the 'O' of (3) are evaluative uses and that the transformation of (1) into (4) and of (3) into (5) reveals a confusion of the evaluative with the prescriptive.[2] Similarly, the outermost 'O' of (2) is used evaluatively, whereas the parenthetic 'O' remains ambiguous. It seems to me that some of the controversy about what are or are not intuitively acceptable deontic truths would be resolved if the two uses of 'ought', although closely connected, were not confounded.

The confusion of interpretation has emerged not only in connection with systems where the deontic operators are taken as primitive but also on some readings of the defined deontic modalities. Alan Anderson,[3] for example, defines

(6) $OA =_{df} (\sim A \dashv 3 \mu)$

where 'OA' is read alternatively as 'It ought to be the case that A' or 'It is obligatory that A', the definiens is taken to be 'If not A then

---

2. In *Normative Discourse* (Englewood Cliffs, N.J.: Prentice-Hall, 1961), Paul W. Taylor elaborates, informally, the distinction between evaluative and prescriptive uses of 'ought'.

3. A. R. Anderson, *The Formal Analysis of Normative Concepts*, Technical Report No. 2; U.S. Office of Naval Research, 1956). More recently, in an unpublished paper, Anderson has replaced (6) with a definition in terms of a weakened implication. The replacement does not resolve the ambiguity in interpretation.

the world will be worse off', and the contrary of 'O' is taken to be 'forbidden' or, equivalently, 'not permitted'. '⊰' is taken to be a strong conditional. The previous dissonance arises as between ought and forbidden. It is clear that an adequate deontic logic must provide for the elimination of redundant iterations and distinguish between ambiguous uses of 'ought' that may be reflected in iterated or nested modalities.

It seems to me that the least problematic reading of the deontic operators is the reading that fits the use of 'obligatory', 'forbidden,' 'permitted', as they occur in connection with rule-governed conduct. Such a restriction to prescriptive use has the advantage of allowing us to be clearer about the semantic interpretation of deontic statements. It is surely apparent that deontic statements are nonextensional, for suppose that, like a well-known comedian, everyone wriggled his eyebrows when and only when he performed a lascivious act. Eyebrow wriggling would be forbidden in an extensional deontic logic. Or consider a Manichean universe where the forces of good and evil are in equilibrium, an act of kindness is always accompanied by an act of cruelty, and conversely. Given extensionality, if cruelty is forbidden, so is kindness. Granted that deontic statements are not truth-functional, when they are embedded in familiar logical systems is it not assumed that they have a truth value? Except for a proposal of Stenius,[4] there has been no suggestion that within the scope of the deontic operators, the logical connectives have altered in meaning. The assumption being made (i.e., that there is no alteration in meaning of words in the logical vocabulary) is plausible where the context is restricted to rule-governed behavior.

It has been suggested by Kurt Baier[5] and others that, given a set of rules and standards, if a given action is roughly covered by those rules and standards, we may take 'It is obligatory that *A*' (where '*A*' is an appropriate statement of the sort '*x* does *w* at *t*') as meaning that *A* is entailed by the set of rules and standards. A forbidden action is described by a statement inconsistent with the code, and so on. As Smiley[6] has suggested,

$$(7) \qquad OA =_{df} (\mu \mathbin{\text{⊰}} A)$$

4. Erik Stenius, "The Principles of a Logic of Normative Systems," *Acta Philosophica Fennica, Proceedings of a Colloquium on Modal and Many Valued Logics* (1963), pp. 247–260.

5. Kurt Baier, *The Moral Point of View* (Ithaca: Cornell University Press, 1958).

6. T. J. Smiley, "The Logical Basis of Ethics," *Acta Philosophica Fennica, Proceedings of a Colloquium on Modal and Many Valued Logics* (1963), pp. 237–246.

lends itself to such an interpretation, where 'μ' stands for a conjunction of statements (categorical declaratives) in a code. In a deontic logic so restricted, redundant modalities ought to be eliminable.

If a deontic logic is also supposed to give an account of the uses of 'ought' as it occurs in (1) and (2) as well, then, without some additional clues as to the semantic interpretation of such statements, the development of deontic systems, however elaborate, may be misleading technical exercises.

3. *Essentialism in Modal Logic*

The source of this paper appeared in the inaugural issue of *Noûs,* I, 1 (March 1967): 90–96. It elaborates on themes about modalities and essentialism sketched in 1961 in "Modalities and Intensional Languages," this volume. Included are occasional stylistic revisions, corrections, and a few clarificatory insertions, but no changes of substance. ■

It has been charged that modal logics are essentialist theories, where an essentialist theory is a theory in which it is possible to distinguish necessary attributes. Talk of necessary attributes is held to be bewildering, senseless, and indefensible.

In this paper I will indicate the manner in which attributes may be introduced in a standard system of quantified modal logic. Let us assume that this system is a quantified S5 with the Barcan formula, where the individual constants are referential names and not singular descriptions (QML). An example of the kind of essentialist argument that is supposed to generate perplexities will be stated in QML. Since the argument is equivocal (as are many informal uses of modal terms), two interpretations of the argument are proposed. The first is shown to be invalid on the given assumptions, and the second, though an instance of a valid form, will not generate categorical necessities of the essentialist sort without the ad hoc addition of clearly essentialist premises. To show the latter requires a more adequate characterization of an essentialist theory (if it is to do justice to the traditional notion).

Attributes in Quantified Modal Logic

In QML the modal operators in well-formed expressions attach to sentences and sentential functions, not to predicates or predicate variables. There is, however, a straightforward derivative sense in which we can speak of necessary (and contingent) attributes. QML includes notation for abstraction[1] and the monadic instance of a generalized axiom schema:

(1) $x \in \hat{y}A \equiv B$ where B is the result of substituting x for every free occurrence of y in A, given the usual constraints

Abstracts may be interpreted as designating attributes. To every proposition about an object a there corresponds one or more attributes assigned to a in that proposition. [See, for example, (18)–(21) below.] Since there are statements of the form '$\Box B$' that hold in QML, there are also statements of the form

(2) $\Box(a \in \hat{x}A)$

---

1. See my ''Identity of Individuals in a Strict Functional Calculus of First Order,'' *Journal of Symbolic Logic*, XII (1947): 12–15. The permutation of ' $\Diamond$ ' and the existential quantifier and its equivalents (the Barcan formula and its converse) are features of QML.

that hold in QML, and we may speak of the attribute in (2) as being "necessary of" or "necessarily assigned to" *a*. Depending on how we understand 'contingent', a contingent attribute may be taken as an attribute such that

$$(3) \qquad \Diamond\,(a\in\hat{x}A)\cdot\Diamond\sim(a\in\hat{x}A)$$

or

$$(4) \qquad (a\in\hat{x}A)\cdot\Diamond\sim(a\in\hat{x}A)$$

The Antiessentialist Argument

The critics of essentialism argue that the talk of necessary attributes leads to bewildering conclusions. W. V. Quine says:

> (5)   Mathematicians may conceivably be said to be necessarily rational and not necessarily two-legged, and cyclists necessarily two-legged and not necessarily rational. But what of an individual who counts among his eccentricities both mathematics and cycling? Is this concrete individual necessarily rational and contingently two-legged or vice versa? Just insofar as we are talking referentially of the object with no special bias toward a background grouping of mathematicians as against cyclists or vice versa, there is no semblance of sense in rating some of his attributes as necessary and others as contingent.[2]

Now suppose we agree that to argue

> (6)   Cyclists are necessarily two-legged. (A dubious claim about physical necessities?)

> (7)   *a* is a cyclist.

hence,

> (8)   *a* is necessarily two-legged.

is bewildering. But this would count against the distinction between necessary and contingent attributes in QML only if arguments like (6)–(8) were valid and if the sort of attribute (two-leggedness) they involve were the sort of attribute that interpretations of QML attribute necessarily to objects. Let us reconstruct the argument in QML. Since

---

2. *Word and Object*, (Cambridge: M.I.T. Press, 1960), pp. 199–200.

there is an ambiguity in the first premise (6), there are two alternatives. On the first alternative, in the language of attributes we have,[3] the argument is

(9)　　$\Box(x)(x \in \hat{y}Cy \supset x \in \hat{y}Ty)$

which is equivalent to $(x)(x \in \hat{y}Cy \exists x \in \hat{y}Ty)$

(10)　　$a \in \hat{y}Cy$

therefore

(11)　　$\Box(a \in \hat{y}Ty)$

But the rule of inference, from $A \ni B$ and A infer $\Box B$, is invalid in QML except where $A \equiv \Box C$, and no such claim of necessity has been made for premise (10).

On the alternative reading we can state the argument as

(12)　　$(x)(x \in \hat{y}Cy \supset \Box(a \in \hat{y}Ty))$

(13)　　$a \in \hat{y}Cy$

therefore

(14)　　$\Box(a \in \hat{y}Ty)$

As it stands, (12)–(14) is valid. But to claim that (14) holds on the hypothesis (12) and (13) is harmless enough. Difficulties would arise if such peculiarly essentialist hypotheses as (12) could be affirmed categorically in QML.

Essentialism Characterized

The traditional essentialist does not claim that all the attributes each object has are necessary to it (although this is not an unknown metaphysical claim—e.g., Leibniz). Second, and more important, the essentialist assumes that not every attribute necessary to an object a is necessary to every object whatever. Implicit in his theory is that there are at least some attributes, e.g., $\hat{y}Ay$, which some objects have necessarily but others may not have at all, or have only contingently. The principles may be stated as follows:

(15)　　There is some attribute $\hat{y}Ay$ such that
　　　　$(x)((x \in \hat{y}Ay) \supset \Box(x \in \hat{y}Ay))$

---

3. See also "Modalities and Intensional Languages," this volume.

(16)   There is some attribute $\hat{y}Ay$ such that
$(\exists x)(\exists z)(\Box(x \in \hat{y}Ay) \cdot \sim\Box(z \in \hat{y}Ay))$

(17)   There is some attribute $\hat{y}Ay$ such that
$(\exists x)(\exists z)(\Box(x \in \hat{y}Ay) \cdot (z \in \hat{y}Ay) \cdot \sim\Box(z \in \hat{y}Ay))$

If (16) holds in a theory, we will describe that theory as weakly essentialist, and if (17) holds, we will describe it as strongly essentialist. Attributes that fulfill (16) or (17) along with (15) will be called essential attributes.

Given the above characterization, we may raise the question with respect to interpretations of QML as follows: Is QML essentialist in either the weak or the strong sense? Also, if it is, are the affirmably essential attributes of QML of the sort considered in (12)–(14)?

Essentialism in QML

QML seems on the face of it to be essentialist in both the weak and the strong sense. As was pointed out above, a proposition in which an object a is referred to may be assigning a plurality of attributes to a. Consider the following proposition:

(18) aIa ('I' for identity)

Three attributes are assigned to a by (18). They are

(19) $\hat{x}(xIx)$

(20) $\hat{x}(aIx)$

(21) $\hat{x}(xIa)$

On the assumption that there are at least two individuals (call them 'a' and 'b'), (20) and (21) fulfill the conditions for being essential attributes under (16) (weak sense), since the following hold in QML:

(22) $\Box(a \in \hat{x}(aIx))$

(23) $\sim\Box(b \in \hat{x}(xIa))$

It is important to note that (19), although necessary, is *not* an essential attribute, since the conditions of (16) are not met. What distinguishes (19) is that it does not mention a. We will say that an attribute is nonreferential with respect to an object a if it is represented by an abstract that does not mention a, e.g., (19). Otherwise it is referential with respect to a. Now (1), the axiom over abstraction, tells us that, although a proposition may be assigning a plurality of attributes to *a* as in (19)–(21), these assignments generate propositions that are strictly equivalent. For example,

(24) $a \in \hat{x}(x \mathrm{I} x) \equiv a \in \hat{x}(a \mathrm{I} x)$

From the substitution principle for strict equivalence it follows that statements containing referential attributes can be replaced in QML by statements containing nonreferential attributes. This suggests that there is an additional sense in which some provably essential attributes are trivial in QML, where a proof carried out with referential attributes may be paralleled by an equivalent proof with nonreferential attributes. No step in the parallel proof depends on its being the particular object that has that corresponding nonreferential attribute, and it may therefore be repeated for every object in the domain. But these are not the bewildering cases of Quine's examples, although they do show, as in (21)–(22), that there are categorical, albeit trivial or vacuous, cases of weakly essential attributes in QML and, as in (27)–(28) below, strongly essential attributes in QML. For, on the assumption that there is in addition to at least two individuals (a,b), at least one predicate *F* such that *Fa, ~Fb,* and, as in the case of an atomic predicate $\sim \Box (Fx)$, it can be shown that QML is strongly essentialist. Consider

(25) $Fb \lor \sim Fb$

Three attributes are assigned to *b* by (25):

(26) $\hat{x}(Fx \lor \sim Fx)$

(27) $\hat{x}(Fx \lor \sim Fb)$

(28) $\hat{x}(Fb \lor \sim Fx)$

It is clear that (27) qualifies as a provably essential attribute in the strong sense, since

(29) $\Box(b \in \hat{x}(Fx \lor \sim Fb))$
        and
        $(a \in \hat{x}(Fx \lor \sim Fb) \cdot \sim \Box(a \in \hat{x}(Fx \lor \sim Fb))$

But, as in the previous case, since any statement that contains a referential attribute can by *n*-adic extensions of (1) be replaced by an equivalent statement that does not contain a referential attribute, such essential attributes are trivialized.

It should be noted that strictly equivalent attributes are not necessarily eliminable if more strongly intensional contexts, such as belief contexts, are allowed, for (1) asserts an equivalence short of identity.

Consider (24), for example. Although a "strict" equivalence holds, there is no isomorphic equivalence of the terms. Indeed, the attribute $\hat{x}(x\mathrm{I}x)$ is true of all objects, whereas $\hat{x}(x\mathrm{I}a)$ is true of only one.

We have claimed that there are some provably essential attributes in QML, where there are at least two members with distinguishable properties in the domain of the interpretation, but these can, so to speak, be analyzed away. The question remains whether attributes of the sort that Quine discusses, e.g., two-leggedness, could be among provably essential attributes of QML. Clearly, they cannot be, for if $\Box(a\epsilon\hat{x}Tx)$ were *categorically* true in QML, since $\hat{x}Tx$ is a nonreferential attribute, any proof of $(a\epsilon\hat{x}Tx)$ could be carried out for any object whatever. But the implicit assumption of traditional essentialism is that such attributes are necessary to some objects but not to all, as in the case of natural kinds.

Although I have argued that QML does not countenance, as essential attributes, attributes of the kind Quine mentions, I am not thereby agreeing that all such talk is senseless. Indeed, Quine's critical remarks in (5) suggest a way in which we might proceed in our analysis. The fact is that the traditional essentialist is *not* "talking referentially of the object with no special bias toward a background grouping. . . ." It is precisely certain kinds of background grouping with which he *is* concerned. Presumably his modalities would be relative to some initial set of *imported* nonlogical premises that affirm a connection between attributes such as those in

(30) $(x)(x\in\hat{y}Ay \supset \Box(x\in\hat{y}By)$

of which (12) is an instance.[4]

---

4. In his dissertation, "The Elimination of Individual Concepts," (Stanford University, 1966), Terence Parsons explores this possibility in some detail. Attributes related as in (30) are called "essence pairs."

4. *Essential Attribution*

A version of this essay was presented on the occasion of a symposium sponsored by the American Philosophical Association and held on December 29, 1970. The other symposiasts were David Kaplan and Saul Kripke. Comments of several colleagues on the symposium paper were helpful, in particular, those of Terence Parsons. The paper was published with some changes in the *Journal of Philosophy*, LXVIII, 7 (April 8, 1971): 187–202.

The previous paper "Essentialism in Modal Logic" attempts a formal characterization of Aristotelian essential properties. This essay introduces, inter alia, another notion of essential property but is centrally concerned with elaborations of Aristotelian themes.

The present printing contains a few corrections and revisions of the published paper of an editorial or clarificatory nature. Also, a few passages that seemed on rereading to be redundant or long-winded have been omitted. The original numbered line (19) has been replaced by the present (19)–(21). ■

I

A sorting of attributes (or properties) as essential or inessential to an object or objects is not wholly a fabrication of metaphysicians. The distinction is frequently *used* by philosophers and nonphilosophers alike without untoward perplexity. Given their vocation, philosophers have also elaborated such use in prolix ways. Accordingly, to proclaim that any such classification of properties is "senseless" and "indefensible," and leads into a "metaphysical jungle of Aristotelian essentialism"[1] is impetuous. It supposes that cases of use that appear coherent can be shown not to be so or, alternatively, that there is an analysis that dispels the distinction and does not rely on equally odious notions. It further supposes that taking the distinction seriously inevitably leads to what W. V. Quine calls "Aristotelian essentialism." The latter claim is a nest of presumptions, two of which are that "Aristotelian essentialism," as characterized by Quine, is a characterization of Aristotelian essentialism, and that any theories that countenance the distinction incorporate versions of "Aristotelian essentialism."

On the occasion where Quine seems to propose an argument against a genuine mode of essentialism, i.e., the case of the mathematical cyclist,[2] it is seen on alternative interpretations that either the argument is invalid or there is no ground for supposing that anyone would accept its premises.[3] Friendly critics suggest that it was not intended as an argument. It was supposed to bewilder us, and it does. But Quine acknowledges that it was never a central purpose to address himself seriously to essentialist claims. Raising specters of essentialism was ancillary to the grander purpose of rallying *further* reasons for rejecting quantified modal logic. The latter, he says, is "committed to essentialism," which in turn is, on the face of it, senseless.

I[4] and Terence Parsons[5] have argued—persuasively I believe—

1. W. V. Quine, *Word and Object* (Cambridge: M.I.T. Press, 1960), pp. 199–200; "Three Grades of Modal Involvement," in *The Ways of Paradox* (New York: Random House, 1966), p. 174.

2. *Word and Object,* p. 199.

3. Shown in "Modalities and Intensional Languages" (1961) and in "Essentialism in Modal Logic" (1967), both in this volume. See also Parsons, Plantinga, and Cartwright, notes 5 and 13 below.

4. See "Modalities and Intensional Languages" and "Essentialism in Modal Logic," this volume. Also "Discussion on the Paper of R. B. Marcus," (1962), in this volume; reprinted in *Boston Studies in Philosophy of Science,* ed. Marx Wartofsky (Dordrecht: Reidel, 1963). See also Parsons, Plantinga, and Cartwright, notes 5 and 13 below.

5. "Grades of Essentialism in Quantified Modal Logic," *Noûs,* 1, 2 (May 1967): 181–191. Also, "Essentialism and Quantified Modal Logic," *Philosophical Review,* LXXVIII, 1 (January 1969): 35–52. For Quine's attempt at formal characterization, see note 23 below.

that Quine's casual characterization of essentialism is inadequate. It misses the point of what seems to be presupposed in coherent cases of use or in Aristotelian essentialism. Furthermore, since Quine's characterization does not take us beyond the distinction between noncontroversial necessary and contingent propositions, there is no cause for perplexity. What *is* perplexing is that his characterization doesn't fit his case of the mathematical cyclist.

On a more adequate characterization, which in the present paper is taken to be minimal, it has been shown[6] that quantified modal logic (QML) is not committed to essentialism (E) in the following sense: in the range of modal systems for which Saul Kripke[7] has provided a semantics, no non-trivial essentialist sentence is a theorem. Furthermore, there are models consistent with all such sentences being false.

Modal logic accommodates essentialist talk. But such talk is commonplace in and out of philosophy.[8] It is surely dubious whether essentialist talk can be replaced by nonessentialist, less "problematic" discourse. The offhand remark that "some attributes count as important and unimportant, . . . some as enduring and others as fleeting; but none as necessary or contingent"[9] (which Quine takes as capturing the distinction between essential and accidental) suggests that such talk is dispensable through some uniform substitution of words that are clearer and untainted by metaphysics. But if Socrates was born and died snub-nosed, then that property and his being a man are equally durable. Are we helped by the further assertion that his being a man is more important? In what way? Is it more important than his being a philosopher? And what are we to make of cases where it is claimed of a certain attribute that it is important but inessential?[10] Is 'important' less problematic than 'essential'?

Given the apparent coherence of some essentialist talk, interpreted

6. Parsons, "Essentialism and Quantified Modal Logic."

7. "Semantical Considerations on Modal Logic," *Acta Philosophica Fennica, Proceedings of a Colloquium in Modal and Many Valued Logics* (1963): pp. 83–94.

8. If the reader is in doubt, he is urged to check the claim against his own reading. Quine himself does not shun such use. It ranges from the metaphorical "Nominalism is in essence perhaps a protest against the transcendent universe" to a precise sorting of a certain property of expressions, the property of occurring in a sentence, as between essential and inessential occurrences. See *The Ways of Paradox*, pp. 69, 73, 103.

9. *Word and Object*, p. 199.

10. The first chapter of M. B. Hesse, *Models and Analogies in Science* (Notre Dame: University of Notre Dame Press, 1966) begins with a discussion between a Campbellian and a Duhemist about whether it is essential to (or an essential property of) a scientific theory that it have a model. The Campbellian claims it is essential. The Duhemist claims it may be important or useful but not essential. Nor is the discussion "without semblance of sense." We recognize in it the same conceptual scheme implicit in such distinctions with respect to properties of more mundane objects.

systems (**I**) of QML are appropriate vehicles for analysis of essentialism. The present paper continues with the account of modes of essentialism within the framework of QML. In particular, it is suggested that *Aristotelian* essentialism may best be understood on a "natural," or "causal," interpretation of the modal operators. But first a few historical remarks.

The early taxonomy of preformal logic distinguished between *pure* and *modal* propositions. According to W. S. Jevons[11] in 1884,

> The pure proposition simply asserts that the predicate does or does not belong to the subject, while the modal proposition states this *cum modo,* or with an intimation of the mode or manner in which the predicate belongs to the subject. The presence of any adverb of time, place, manner, degree, etc., or any expression equivalent to an adverb, confers modality on a proposition. "Error is always in haste," "justice is ever equal," . . . are examples of modal propositions.

Further on Jevons mentions that some logicians have adopted a special view with respect to 'necessarily', 'possibly', 'generally', and the like, as in "an equilateral triangle is *necessarily* equiangular," "men are *generally* trustworthy," where the "modality does not affect the copula of the proposition" but "consists in the degree of certainty with which a judgment is made or asserted."

With formalization came the faith that standard functional logic (SFL), appropriately interpreted, would yield an analysis and disambiguation of modal propositions. Logical grammar would replace surface grammar, yet the *sense* (if it had one at all) of the original would be *captured* by its formal, nonmodal counterpart. The faith was not ungrounded. Jevons used 'always' and 'ever' as examples. Interpretations (**I**) of SFL take us a long way. The adverb is detached from the predicate. The nontemporal 'always' and 'sometimes' go into quantification. The temporal cases go into **I** of SFL, which includes temporal moments in the domain of **I**. Similarly for the nonspatial (rare) and spatial 'everywhere' and 'somewhere'. But even in the temporal cases the success is incomplete. The modalities proved more recalcitrant, and extensions of SFL proved useful. Those who frown on such extensions as deviant should remember Rudolf Carnap's admonition: "In logic there are no morals."[12]

11. *Lessons in Logic* (London and New York: Macmillan, 1884), pp. 69–70. Jevons seems to be making a *de re, de dicto* distinction in cases where the modality "consists in the degree of certainty with which a judgment is made."
12. *The Logical Syntax of Language* (New York: Harcourt, Brace, 1937), p. 52.

II

I should like to focus on two modes of essentialism, which I will distinguish as *individuating* and *Aristotelian*. Consider some cases. We say of Moby Dick that although he lives in the sea he is essentially a mammal, and of Socrates that he is essentially a man and accidentally snub-nosed. We point to a sample of mercury at room temperature and say that, although it is a liquid, it is essentially a metal, suggesting that solidity at room temperature is an accidental property of metals. These are cases that fit Aristotle's account of essences. What is implicit here? The objects are actual objects, and the properties that are being *sorted* as essential or inessential correspond to direct, nonvacuous, "natural" predicates. (A more formal but inevitably approximate characterization of such predicates is deferred for subsequent discussion.) For Aristotelian essentialism an essential property is a property that an object *must* have. It answers to the question "What is it?" in a strong sense; if it ceased to have that property it would cease to exist. It is a nonuniversal property such that, if anything has it at all, it has it necessarily. The latter condition is what distinguishes Aristotelian from what I call "partially individuating" essentialism.

Consider, for example, Winston[13] the mathematical cyclist. Suppose he is an avid cycling enthusiast. Cycling is an overriding preoccupation. Although he holds a position on a mathematics faculty, his interest in that subject is at best desultory. Arguing against a renewal of Winston's contract, a colleague says, "Unlike the rest of us, Winston is essentially a cyclist, not a mathematician." Analogously, Protagoras might have said of Socrates, "He's essentially a philosopher, not a politician." The social worker says of a client, "He's essentially a good boy; just fell in with bad company," which is after all not too distant from distinguishing, as philosophers sometimes do, those who are disposed in character to act rightly from those who merely act rightly out of expediency or the like.

Implicit in such examples of partial individuation is that among the attributes an object must have are not only those that it shares with objects of its kind (Aristotelian essentialism) but those that are *partially* definitive of the special character of the individual and distinguish it from some objects of the same kind. But must there be some set of individuating essential attributes that *wholly* distinguish an ob-

13. So named by R. Cartwright in "Some Remarks on Essentialism," *Journal of Philosophy*, LXV, 20 (October 24, 1968): 615–626, p. 619. Also called "Squiers" by Alvin Plantinga in *"De Re et De Dicto,"* *Noûs*, III, 3 (September 1969): 235–258.

ject from those of its kind? (For present purposes, we need not go into the question of the uniqueness of proximate kinds, the hierarchy of kinds, and the like.)

Being a snub-nosed, henpecked, hemlock-drinking philosopher[14] wholly individuates Socrates without the addition of a uniqueness condition, but, although being a philosopher may be essential to his particular nature, presumably being snub-nosed is not. Perhaps complete individuation is always a matter of what are generally taken to be inessential properties, accidents of circumstance. If we encountered a winged horse, we could not determine that it was Pegasus unless we knew some of the circumstances of its birth and history. But then, those circumstances could be inessential.

Numbers, as contrasted with empirical objects, are supposed to be objects that can be wholly individuated by their essential properties, without tacking on a uniqueness condition. But the matter is by no means clear. Consider, for example, the counterclaim that numbers have no essential properties at all, for if they had, "it would conflict with the idea that number theory can be reduced to set theory in various ways."[15] One possible resolution to this disagreement is that, if some-

14. The example is from Daniel Bennett, "Essential Properties," *Journal of Philosophy,* LXVI, 15 (August 7, 1969): 487–499, p. 487.

15. Gilbert H. Harman, "A Nonessential Property," *Journal of Philosophy,* LXVII, 6 (March 26, 1970): 183–185. Harman's claim raises interesting questions. If one supposes that numbers are first-order objects, that there is at most one of each, and that there are nevertheless equally acceptable alternative choices for the natural-number structure, then any "world" W that includes natural numbers in its domain D does so by *specifying* which object it identifies with 0, which with 1, and so on. Although there may be other subsets of D with structures (N_1, S_1, d_1) isomorphic to the given choice for (N, S, d), where N is the set of number elements, S the successor relation, and d the distinguished element, elements of N_1 that are isomorphic to elements of N will not be *the* numbers in that world W. See also P. Benacerraf, "What Numbers Could Not Be," *Philosophical Review,* LXXIV (1965): 47–63.

The truth of Harman's claim, i.e., that numbers have no essential properties at all, revolves about what is meant by "reducing number theory to set theory." If it means that there is no such thing as *the* natural-number structure, but only some set of alternative isomorphic structures (N', S', d'), one for each W', each of which represents some alternative set-theoretic reduction, then his claim is correct. And, as T. Parsons has shown in "Essentialism and Quantified Modal Logic," pp. 44–46, QML may be extended to include arithmetic truths without the consequence that numbers have any essential properties at all. '9 is necessarily greater than 7', for example, comes to: 'In each possible world there is something that is 9 and something that is 7 such that the first is greater than the second'. Furthermore, 'is the number of planets' is not substitutable for 'is 9' in accordance with principles of substitution in modal contexts.

But if one takes it that there is something that is *the* natural-number structure *tout court,* as Georg Kreisel suggests in "Mathematical Logic: What Has It Done for the Philosophy of Mathematics?" in *Bertrand Russell, Philosopher of the Century,* ed. R. Schoenman (London: Allen & Unwin, 1967), pp. 213–216, then alternative reductions are isomorphic to but not identical with the series of natural numbers.

There is of course a deficiency in the way Harman presents his thesis. He seems to

thing counts as a number, it has essential numerical properties, but they do not wholly individuate.

Perhaps with respect to inquiries like "Who is Sylvia? What is she?" the latter question can be answered in terms of essential properties (Aristotelian), but individuating *essences* can never *wholly* answer the former. This leads some philosophers to make a metaphysical shift. They invent *objects* (individual concepts, forms, substances) called "essences," which have only essential properties, and then worry when they can't locate these objects by rummaging around in other possible worlds. It does not seem to me that an account of essential attribution compels us, even with respect to abstract objects, to shift our ontology to individual essences. The usefulness of talk about possible worlds is not for purposes of individuating the object—that can be done in this world; such talk is a way of sorting its properties.

Within the possible-world view of QML the matter may be put as follows: among all the direct, nonvacuous, "natural" properties [16] an object has in this world (W), there are those it must have. Among those it must have are those it has in common with objects of some proximate kind (Aristotelian essentialism) and those that partially individuate it from objects of the same kind (individuating essentialism). We see that to say of an object x and an object y that they have all essential properties in common is weaker than claiming identity. But this reflects common speech. To say of x and y that they are essentially the same (the same in essential respects) is a weaker claim than saying they are identical.

What has gone wrong in recent discussions of essentialism is the

---

suggest that one can talk of the properties of numbers, independent of their being part of the structure that makes them eligible for numberhood. This is analogous to the specious arguments that might develop about whether some carved piece of wood was the queen in a chess game.

16. In what follows, by indicating how we "give" an interpretation for purposes of paraphrase (as distinguished from specifying how we make truth assignments to sentences of our language) and by placing certain restrictions on predicates, we have excluded many predicates that Hume and others would have called nonnatural. See Hilary Putnam, "The Thesis That Mathematics Is Logic," in *Bertrand Russell, Philosopher of the Century,* ed. R. Schoenman, (London: Allen & Unwin, 1967), pp. 299–301, for a brief discussion of the "natural" and "philosophical" notion of a predicate or property. "Natural" is roughly contrasted with "artificial."

Carnap, in *The Logical Syntax of Language,* pp. 308–309, recommends exclusion from the class of property-words those that correspond to what he called "transposed properties." From his examples we see that his notion of "transposed properties" overlaps the loose traditional characterization of "nonnatural properties." Among his examples is the property a city has if its name is the alphabetical predecessor of the name of a city with more than 10,000 inhabitants. He also includes properties like being famous, or being discussed in a certain lecture, as not being "qualities in the ordinary sense."

assumption of surface synonymy between 'is essentially' and *de re* occurrences of 'is necessarily'.[17] But intersubstitution often fails to preserve sense. Would Winston's colleague have been as well under- stood if he had said "Winston is necessarily a cyclist?" And would we ever be inclined to use 'is essentially' instead of 'is necessarily' where vacuous properties are concerned, as in 'Socrates is essentially snub-nosed or not snub-nosed' or 'Socrates is essentially Socrates'? If higher-order objects are candidates for essential attribution (as R. Cartwright freely allows) then substitution will sometimes take us from truth to falsity, as in '*p* is essentially correct'. The connection between the two locutions is not a surface matter. It is best analyzed within some model of QML.

III

Which of the Kripke model structures (W, K, R) are suitable for our analysis? T. Parsons's results clearly exclude maximal models (where R is symmetric and transitive as well as reflexive). Intuitive consid- erations suggest that, so far as Aristotelian essentialism is concerned, the chosen system should perhaps satisfy the converse of the Barcan formula (the domain of this world W includes the domain of each K' possible relative to W) and R should be transitive (i.e., quantified S4). But that is merely a suggestion. Nothing we are claiming in *this* paper rests on that particular choice of a system.

Whenever our purpose is to use an interpreted formal language for paraphrase and analysis of an ordinary sentence, how we specify the interpretation **I** is crucial, if, in addition to preserving truth in trans- lation, we want somehow to preserve meaning to the maximum extent. It is perhaps gratuitous to emphasize that this is true of interpretations of standard functional logic (SFL), as it is of QML. We would not, in an **I** of SFL (which includes numbers in its domain), choose, from among all possible "names" for 9, 'the number of planets', for then the ordinary sentence 'the number of planets is (identically) 9' would go into **I** as 'the number of planets is the number of planets'. We would not make that choice because we are not inclined to obliterate mean- ings unnecessarily. Nor would we choose 'the henpecked, snub-nosed, hemlock-drinking philosopher' over 'Socrates' as the "name" we as- sociate with some constant for designating that individual.

Such considerations are crucial for translatability into QML, for

17. See for example, Plantinga, *"De Re et De Dicto"*; also "World and Essence," *Philosophical Review*, LXXIX, 4 (October 1970): 461–492.

the stratagem of talk about possible worlds is that truth assignments of sentences and extensions of predicates may vary, but individual (nondescriptive) proper names don't alter their reference, except to the extent that in some worlds they may not refer at all. If, therefore, we take as Socrates' name a singular description that picks him out in this world only, our purpose is defeated at the outset. What we want is that neutral peg on which to hang descriptions across possible worlds. Similarly, the "sense" of sentences and predicates is preserved across possible worlds. For those who are quick to argue that *ordinary* names cannot *always* be used in such a purely referential way, we can, in giving the interpretation, expand our lexicon to provide neutral, purely referential names where necessary.

Given some choice of an appropriate system, we specify as follows: Associated with sentence symbols are ordinary sentences (not descriptions of sentences). Associated with individual constants are ordinary names, not singular descriptions. Where ordinary proper names are lacking, such nameless objects are first given "ordinary" names by a suitable convention for avoiding duplication of names. (A lexicon is kept.) Where more than one object has the same name, we distinguish them by a suitable convention. For symmetry we might add that, when one object has several names, we choose one as its standard name. This would be required for contexts that have greater obliquity than the modal contexts here considered. We are restricting "indirect" allowable occurrences of variables to those that occur within the scope of a modal operator so interpreted as to permit intersubstitutability *salve veritate* of names (not descriptions) of the same object.

Associated with predicate symbols are standardized predicates. A standardized predicate is like an ordinary sentence modified as follows: disambiguate names of multiple reference. Replace one or more occurrences of names by place markers. Quine's "standard English predicate," for example, will serve with the following extension: the only indirect occurrences of names replaceable by place markers are those that fall within the scope of a modality translatable into QML. Suppose, for example, as Leonard Linsky claims, that "the statement 'I did not miss this morning's lecture, but I might have' and 'I did not miss this morning's lecture, but there is a possible world in which I did' are full paraphrases of each other." [18] Then, since 'might' goes

18. "Reference, Essentialism, and Modality," *Journal of Philosophy*, LXVI, 20 (October 16, 1969): 687–700. We are excluding here epistemic contexts along with stronger obliquity in the formation of predicates. For example, 'John' and 'Jill' may both be replaced in 'John might have married Jill', but only 'John' may be replaced in 'John knew Jill left town' or 'John wished Jill would marry him'.

into '◇', the predicate formed from Linsky's sentence is indirect, i.e., contains an allowable indirect occurrence of a place marker.

There remains the representation of singular descriptions. If our choice of (W, K, R) contains identity, then the theory of descriptions with attention to scope will work. An alternative is taking singular descriptions (not proper names) as uniquely satisfiable predicates.[19]

A word of caution here. In specifying how we paraphrase, we hope to avoid a few confusions. Plantinga, for example, has staked several arguments on the claim that being snub-nosed in W is a property Socrates has in all possible worlds that contain him and is, therefore, essential. Are we to suppose that 'Socrates is snub-nosed in W' (*P*), like 'Socrates was born in Athens', is one of those ordinary sentences we associate with sentence symbols of our interpreted QML; that in the domain of our interpretation there are places, one of which is Athens and another the world? All that Plantinga's funny sentence might come to is that, in our choice of (W, K, R), its truth assignment is T in W and so, therefore, must be the assignment to $\Diamond P$. If we should also choose QS4 as our basis, it will not follow that $\Box \Diamond P$ is assigned T.[20] Furthermore, if we are to conform to coherent cases, we will argue below that indirect predicates are excluded from the characterization of essentialism, for would anybody, essentialist or no, ever want to say that Socrates is essentially possibly snub-nosed?

Implicit in Aristotelian essentialism is that an object has attributes necessarily that are not necessary to other objects. To say that Socrates is essentially a man is to take as true

$$(1) \qquad \Box F(s) \cdot (\exists x) \sim \Box F(x)$$

from which it follows that

$$(2) \qquad (\exists x) \Box F(x) \cdot (\exists x) \sim \Box F(x) \tag{E_M}$$

Indeed, (2) may be taken as characterizing minimal essentialism, where *F* is any monadic predicate that contains no individual constants. (2) excludes such tautological predicates as $F(x) \lor \sim F(x)$. There is another kind of quasi-vacuous predicate, the partial instantiation of

---

19. In "Modalities and Intensional Languages," (this volume), Russell's theory or, alternatively, taking descriptions as unit attributes or unit properties was proposed. In either case merely extensionally equivalent expressions (sentences in the theory of descriptions, predicates in the other) are not intersubstitutable in modal contexts.

I have omitted here any proposals for reinterpretation of quantification, along substitutional lines. It is important to separate the grounds for such a view of quantification from its usefulness in connection with substitution in indirect contexts. We are, however, presuming a difference between names and descriptions.

20. We have not presumed symmetry of the alternativeness relation.

a tautological predicate, that the essentialist does not count as designating essential attributes; e.g., $F(s)\lor \sim F(x)$ which is necessarily true of s. In order to exclude such cases, as well as for extending our characterization to relational attributes, T. Parsons (see note 5) has generalized (2) to F^n in such a way as to sort out those cases. For simplicity of presentation I will restrict the discussion to monadic predicates that are general (i.e., contain no individual constants).

To those who argue that the exclusion of vacuous predicates is arbitrary and *post hoc* we need only point out that Aristotle excluded them, for such philosophers also claim that they have in mind some version of Aristotelian essentialism. Bennett (see note 14 above) sums up the Aristotelian view as follows:

> Being an entity is a necessary property of everything, i.e., a transcendental property. . . . Essential properties sort the entities of which they are true in some fashion. (p. 487)

> Being an entity, like being self-identical and being a unity, failed to sort Socrates from anything. Everything is an entity, self-identical, a unity. Being identical with Socrates, on the other hand, sorted Socrates from everything. Nothing but Socrates is identical with Socrates. Essential properties are not transcendental, and they are not . . . individuative. (p. 494)

As we noted above, an extension of QML that includes truths like (2) supposes that there is more than one thing and that there are necessities other than logical. But this conforms to cases. Aristotle, in his theory of essences, was after all concerned with some kind of natural necessity. Indeed, if it is true, as some claim, that numbers have essential properties (that meet the condition of minimal essentialism), did not Kant classify such truths as synthetic, although a priori?

We now see that, although it is mistaken to claim that "*any* quantified modal logic is bound to show favoritism among the traits of an object" and "must settle for essentialism,"[21] this *is* true of a QML that imports even minimal essentialism, for (2) *does* show favoritism among the traits of an object. Surely if I were to say of someone that he was necessarily either snub-nosed or not snub-nosed, he could not accuse me of playing favorites among *his* special traits. But if I were to say that he is essentially a man or essentially a philosopher, that *is* playing favorites; but is it on the face of it, senseless? If a Quinean critic still finds it so, he is not thereby compelled to reject QML al-

---

21. W. V. Quine, *From a Logical Point of View*, rev. ed. (Cambridge: Harvard University Press, 1961; New York: Harper & Row, 1963), p. 155.

together, for he can restrict himself to maximal models of QML, in which essentialist statements are taken as false. The ontology of such models is one of bare particulars.[22]

IV

In the present section I will discuss extensions of (2), which fit some modes of essentialism.

(A) Where an individual has essential attributes it is usually implicit that it has attributes that are not necessary.[23] This could be represented as

$$(3) \quad (\exists x)(\Box F(x) \cdot G(x) \cdot {\sim} \Box G(x)) \cdot (\exists x) {\sim} \Box F(x) \tag{E*}$$

One might want to weaken E* by substituting '$\Diamond G(x)$' for the first occurrence of '$G(x)$' or further strengthening the second conjunct. But our interest is not in spinning out alternatives. What is worth indicating is that, if E* is presumed, we see why statements like '9 is essentially composite' may strike us as odd if we believe that all the attributes of a number are necessary. However, against a background of set-theoretic reductions, one might want to claim (for a world with a von Neumann reduction) that being less than its successor is an essential property of numbers and being a member of its successor is not.

But, as Harman ("A Nonessential Property") has pointed out, there are difficulties here. On a Russellian account of numbers, E* has greater plausibility. If our characterization is extended to higher-order objects, then it is a nonnecessary property of 9 that it is a property of the property of being a planet. The extension of the latter varies across worlds. The number of planets might be 8 in some world that is possible relative to W.

(B) Individuating essentialism. An extension of E_M that conforms to cases of what I have called "partially individuating essentialism" is

$$(4) \quad (\exists x) \Box F(x) \cdot (\exists x)(F(x) \cdot {\sim} \Box F(x)) \tag{E_I}$$

22. See notes 4–6 above.

23. On the one occasion where Quine attempts a formal characterization, he chooses for "Aristotelian essentialism," the first conjunct of (3); see "Three Grades of Modal Involvement," p. 174.

(C) Aristotelian essentialism. In contrast with E_I, Aristotelian essentialism takes it that, if anything is a man or a mammal, it is so necessarily. These are not properties that *anything* can have *per accidens*. The same strong condition extends to properties (e.g., rational-animal) that are definitive of a kind (e.g., man). Versions of this condition on an attribute F are

(5) $\quad (x)(F(x) \supset \Box F(x))$

(6) $\quad (x)\Box(F(x) \supset \Box F(x))$

(7) $\quad \Box(x)(F(x) \supset \Box F(x))$[24]

Conjunction of one of (5)–(6) with E_M or E* will give us some mode of Aristotelian essentialism. Since Aristotle did seem to presume accidental attributes, conjunction with E* is plausible, as in

(8) $\quad (x)(F(x) \supset \Box F(x)) \cdot (\exists x)(\Box F(x) \cdot G(x) \cdot {\sim}\Box G(x))$
$\quad \quad \cdot (\exists x){\sim}\Box F(x)$ $\hfill$ (E*$_A$)

Attributes satisfying some mode of Aristotelian essentialism are seen to be disjoint with what we call partially individuating essences.

E_I is a perplexing thesis, although suggested by use. However, we can take (4) and (8) (or one of the alternatives to (8)) as sorting non-Aristotelian from Aristotelian essential attributes.

(D) Further modifications. As we noted above, a further restriction on eligible predicates, in addition to generality, is that they should be initially nonmodal, i.e., direct predicates. Paraphrasing Jevons, the essentialist is not in the first instance intimating the mode or manner of the mode or manner in which the predicate belongs to the subject. Indeed, iterated adverbs, modification of modifiers, are infrequent in ordinary discourse, although within the semantics of QML we can make sense of iterated modalities.

If, in addition to generality and directness, which are required for conformity to cases of essential attribution, we further restrict our predicates with some loose approximation of "natural" predication in mind (N-predicates), an interesting result follows with respect to the relation between *de re* and *de dicto* modalities. In our formal lan-

---

24. If only the converse of the Barcan formula holds, (7) is not equivalent to (6). We have omitted discussion of an important question as to how, if we were to import such truths, to apply the rule of necessitation.

guage we can form "artificial" predicates[25] in the very imprecise sense that translation back into colloquial speech ranges from extremely awkward to impossible. Consider the predicate formed from so simple a sentence as 'Someone offered Socrates poison'. To what property of Socrates does it correspond? Is it the property of being offered poison by someone? Yes, but we can also see that predicates that contain quantifiers even when they are just beyond minimal complexity border on the inexpressible. The same is true of predicates that contain sentence parts. What property does an object have if it satisfies 'All ravens are black $\cdot x = x$'? If we take as our stock of eligible E-predicates those that have no quantifiers and no sentence parts, and are direct and general (N-predicates), then, as T. Parsons[26] has shown, if E_M is true where F is as above, then, for any nonmodal sentence S, if S is not already a theorem, $\Box S$ is not entailed by E_M. Those, like Plantinga, who imagine that with sufficient cunning they can "reduce" the essentialist's *de re* truths to *de dicto* truths have not been sufficiently attentive to these results.

In our specification of N-predicates we could of course simply have required that they be built up out of atomic predicates and truth-functional sentential connectives. But that *would* have appeared *post hoc*. Our purpose was to frame these restrictions within the context of reasons for accepting them.

Here, as elsewhere in this paper, when we say we are "characterizing" Aristotelian essentialism and the like, we are not suggesting that such characterizations are complete. There is good reason to believe that a complete characterization of Aristotelian essentialism (if it is possible) would further require the introduction of temporal modalities, for otherwise, how would we say of an object that when it ceased to have its Aristotelian essence it would cease to exist altogether? Similarly, we are not supposing that our N-properties fully correspond to natural properties (if there is such a characterization), for 'being a number or else a philosopher and a cow' would then count as an N-predicate. Further specification would require, perhaps among extensions of QML with meaning postulates, something like a theory of categories. Still, the predicate 'being a number or else a philosopher and a cow' is at least intelligibly expressible.

<hr>

25. Parsons's requirement of generality simplified the characterization of essentialism for nth-degree cases. With an abstraction operator (as in my "Essentialism in Modal Logic" and in his "Grades of Essentialism in Quantified Modal Logic"), the ultimate generality of the predicate can be preserved; e.g., '$a \in x/x = x$' can be transformed into '$<a,a> \in xy/x = y$'. The generality requirement in addition fits some loose notion of "natural" predicate.

26. "Essentialism and Quantified Modal Logic," pp. 47–48.

Let us return now to the question of the "commitment" of QML to essentialism. For Kripke it was sufficient to define a model in terms of a set of assignments of truth values to sentences, extensions to predicates, along with specification of the domain (a subset of the union of domains of members of K) for each K′ in K. For any model short of a maximal model, there will be some object and some P-assignment such that, in any K in which the object exists, it will be the extension of P. In this sense, we might say that, for nonmaximal models, QML is "committed" to essentialism. But, as I have suggested, essentialist talk is frequently unproblematic. With careful specification of how we paraphrase such talk in QML, we can characterize some modes of essentialism. And, as I will claim for at least one mode, Aristotelian essentialism as here characterized, it is firmly entrenched in the logic of causal statements.[27]

V

We see that modes of essentialism can be characterized in interpretations of QML. I want to suggest that Aristotelian essentialism as here characterized is perhaps best understood where the modalities are taken to be causal or physical modalities.

Consider some familiar examples. I say of a sample (s) that if I put it in aqua regia (R) it would dissolve (D). We do not take such a claim to be unintelligible. Suppose we interpret '$\Box$' of our QML as causal or natural necessity. Then our example may be represented as

(9) $\Box(R(s) \supset D(s))$

Suppose I say of another sample of some different material (u) that if I immersed u in aqua regia it would not dissolve; from this it would follow that

(10) $\sim\Box(R(u) \supset D(u))$

and, therefore, that

(11) $(\exists x)\Box(R(x) \supset D(x)) \cdot (\exists x) \sim\Box(R(x) \supset D(x))$

which is an instance of minimal essentialism.

We may think of s as having the essential attribute of dissolving

27. Recent discussions of essentialism have focused on numerical statements and the like. T. Parsons showed that QML can be extended to include meaning postulates and arithmetic truths in such a way as wholly to avoid any essentialist consequence. Furthermore, there is an intuitive plausibility to those nonessentialist alternatives for construing analytic statements in the broad sense of 'analytic'. On a "natural or causal" interpretation of the modalities, essentialism does not appear to be avoidable, and it is welcomed.

when immersed in aqua regia, in all worlds that are causally possible relative to W and where *s* exists.

Shylock tells us that if you prick him he will bleed and if you tickle him he will laugh, and if you poison him he will die, and if you wrong him he will take revenge. Some of Shylock's assertions could be represented as in (9), e.g., if you poisoned him he will die. If he had added, the same is not true of a stone, then we would have explicit essentialism, and it is all perfectly coherent. Let us suppose (9) is true and someone were to ask, "Why?" Why would *s* dissolve if it was immersed in aqua regia? Why would Shylock bleed if pricked, die if poisoned? An appropriate answer to the first question is "Because *s* is gold." To the second, if the question is not divided, "Because Shylock is a human being." The answers to these questions begin with 'because', but what follows is not an event description but a statement that attributes a kind property to the object. How is it that an object's being of a certain *kind* is a ground (apparently causal) for its having some further essential property or for there being some causal or necessary connection between its nonessential properties, e.g., between the pairs (being immersed in aqua regia, dissolving), (being poisoned, dying)?

Define the corresponding causal conditional as follows:

(12) $S \rightarrow_c P =_{df} \Box(S \supset P)$

then it must be that (9) follows from

(13) $(G(s) \cdot R(s)) \rightarrow_c D(s)$

which in turn instantiates some general law. But which one?

There are alternatives, since the modal operator may be inside or outside the quantifier. If our QML choice is QS4 with the converse of the Barcan formula, then the weaker alternative is

(14) $(x)((G(x) \cdot R(x)) \rightarrow_c D(x))$

The difference between (14) and other versions is illuminating with respect to questions of invariance of laws, but we will defer such considerations. Our question now is: How do we get from (13) to (9), which we can rewrite as

(15) $R(s) \rightarrow_c D(s)$

in QML? Restricted exportation[28] on (13) gives us

(16) $G(s) \rightarrow_c (R(s) \supset D(s))$

---

28. Failure of unrestricted exportation is a characteristic of conditionals such as strict implication and Anderson and Belnap's entailment, which are stronger than the material

This is where Aristotelian essentialism comes in. Being gold or being a human being is not accidental. Then they must conform to one of (5)–(7). Therefore, from $G(s)$ we get $\Box G(s)$, which, together with the modal principle $(P\rightarrow_c Q)\rightarrow(\Box P\rightarrow_c\Box Q)$ and (16), gives us (15).

No metaphysical mysteries. Such essences are dispositional properties of a very special kind: if an object had such a property and ceased to have it, it would have ceased to exist or it would have changed into something else. If by bombardment a sample of gold was transmuted into lead, its structure would have been so altered and the causal connections between its transient properties that previously obtained would so have changed, that we would not reidentify it as the same thing.

Suppose, for comparison, $G(s)$ has been exported in (13) instead of $R(s)$,

$$(17) \quad R(s)\rightarrow_c(G(s)\supset D(s))$$

Since R is not one of those special dispositional kind properties, no causal connection obtains between being gold and dissolving; simply being gold would not count as a cause of dissolution any more than simply being human would alone be a cause of bleeding. On the other hand, although immersion in aqua regia ($R(s)$) is an accidental property, *being* aqua regia is not. The more general law would be instantiated as

$$(18) \quad (A(a)\cdot G(s)\cdot R'(sa))\rightarrow_c D(s)$$

where '$R'(xy)$' is read as "x is immersed in y." And if a were a sample of aqua regia and s a sample of gold, then it would follow that if s were immersed in a then s would dissolve, for, analogously, on restricted exportation,

$$(19) \quad (A(a)\cdot G(s))\rightarrow_c(R'(sa)\supset Ds)$$

But since A and G are kind properties

$$(20) \quad \Box A(a) \text{ and } \Box G(s)$$

they conform to one of (5)–(7), and so it follows that

$$(21) \quad R'(sa)\rightarrow_c Ds$$

Instead of leading into a metaphysical jungle, formulation of our analysis within a causal interpretation of QML is suggestive and il-

---

conditional. The causal analogue of an unrestricted deduction theorem is surely counterintuitive. See the appendix, this chapter.

luminating. Different modes of essentialism will place different interpretations on the modalities. But, so far as a causal interpretation goes,
it suggests interesting relations between laws like (14) and the paradox
of confirmation; it allows for solutions to problems about substitution
in causal contexts; it raises interesting questions about the generality
of laws.

APPENDIX 4A: STRICT IMPLICATION, DEDUCIBILITY, AND THE DEDUCTION THEOREM

It seems appropriate to include this paper (*Journal of Symbolic Logic,* XVIII [1953]: 234–236) since its findings are specifically alluded to in the concluding section of "Essential Attribution." Rather than viewing the restrictions on a deduction theorem as a failing, it would appear that for "intensional" conditionals, those restrictions make for a better fit. ■

Lewis and Langford[1] state, "It appears that the relation of strict implication expresses precisely that relation which holds when valid deduction is possible. It fails to hold when valid deduction is not possible. In that sense, the system of strict implication may be said to provide that canon and critique of deductive inference which is the desideratum of logical investigation." Neglecting for the present other possible criticisms of this assertion, it is plausible to maintain that if strict implication is intended to systematize the familiar concept of deducibility or entailment, some form of the deduction theorem should hold for it. The purpose of this paper is to analyze and extend some results previously established[2] that bear on the problem.

We will begin with a rough statement of some relevant considerations. Let the system S contain among its connectives an implication connective 'I' and a conjunction connective '&'. Let A_1, A_2, . . . , $A_n \vdash B$ abbreviate that B is provable on the hypotheses A_1, A_2, . . . , A_n for a suitable definition of "proof on hypotheses," where A_1, A_2, . . . , A_n, B are well-formed expressions of S.

Three deduction theorems for S would be

1. C. I. Lewis and C. H. Langford, *Symbolic Logic* (New York and London: Methuen, 1932), p. 247.

2. See my "Deduction Theorem in a Functional Calculus of First Order Based on Strict Implication," *Journal of Symbolic Logic,* XI (1946): 115–118. The results of this paper were obtained for functional extensions of the Lewis systems S2 and S4, where S2 and S4 are exactly as formalized by Lewis. The results also obtain for the functional extensions of the Lewis systems under discussion.

I.   If $A_1, A_2, \ldots, A_n \vdash B$ then $A_1, A_2, \ldots, A_{n-1} \vdash A_n \, I \, B$.

II.   If $A_1, A_2, \ldots, A_n \vdash B$ then $(A_1 \, \& \, A_2 \, \& \, \ldots \, \& \, A_n) \, I \, B$ is a theorem.

III.   If $A_1 \vdash B$ then $A_1 \, I \, B$ is a theorem.

It is well known that, for the system of material implication, a meta-theorem corresponding to I and, consequently, to II and III is provable. For the Lewis systems[3] this is not the case for the strict conditional. It was established[4] that, for S2, no theorem is available corresponding to either I, II, or III. More explicitly (since the Lewis systems involve two implication connectives), neither of the following is provable in S2:

(1)   If $A_1, A_2, \ldots, A_n \vdash B$ then $A_1, A_2, \ldots, A_{n-1} \vdash A_n \supset B$.

(2)   If $A_1, A_2, \ldots, A_n \vdash B$ then $A_1, A_2, \ldots, A_{n-1} \vdash A_n \dashv 3 \, B$.

Since S2 is an extension of S1, the above establishes the nonprovability of a deduction theorem for both of these systems. It appears to the author that, in view of Lewis and Langford's proposed analysis, a theorem corresponding to III should (at the least) be available.

For S4, it was established that (1) holds, and (2) holds if the following condition is satisfied.[5]

(3) $A_1 \equiv \sim \lozenge \sim H_1, \, A_2 \equiv \sim \lozenge \sim H_2, \, \ldots, \, A_n \equiv \sim \lozenge \sim H_n$

The above proof may also be paralleled for S5, which is an extension of S4. That (2) does not hold unconditionally for S4 and S5 can be shown by a matrix of Wajsberg's[6] that satisfies the axioms and rules of S4 and S5. Although $A, A \supset B \vdash B$ can be shown in S4 and S5, $A \dashv 3 \, ((A \supset B) \dashv 3 \, B)$ does not always have a designated value.

These results as they stand are perhaps misleading. From them Paul Rosenbloom[7] concludes that "the contention that from the standpoint of the interpretation as deducibility, 'strict' implication is a more satisfactory operation than 'material' implication is consequently untenable until a system based on the former is constructed in which the deduction theorem [for the strict conditional] is proved. . . ." In our previous paper, it was not made explicit whether theorems corre-

---

3. See Lewis and Langford, *Symbolic Logic*, particularly Appendix II.

4. For this and subsequent references to what has been established with respect to the deduction theorem in S2 and S4, see my "Deduction Theorem in a Functional Calculus of First Order Based on Strict Implication."

5. As suggested originally by the referee, this condition could have been weakened to $A_1 \equiv \sim \lozenge \sim H_1, \, A_2 \equiv \sim \lozenge \sim H_2, \, \ldots, \, A_{n-1} \equiv \sim \lozenge \sim H_{n-1}$.

6. See Lewis and Langford, *Symbolic Logic*, the Group III matrix on p. 493.

7. *The Elements of Mathematical Logic* (New York: Dover, 1950), p. 60.

sponding to II or III were provable for S4 (and consequently S5), which indeed they are.[8] This is established by using the following, which are provable for S4:

(4) If A is provable then $\sim \diamond \sim A$ is provable.[9]

(5) $(A_1 \supset \ldots (A_{n-1} \supset (A_n \supset B) \ldots)) \equiv ((A_1 \bullet \ldots \bullet A_n) \supset B)$

From 1 and 4 we have

(6) If $A \vdash B$ then $A \dashv 3 B$ is a theorem.

From 1, 4, and 5 we have

(7) If $A_1, A_2, \ldots, A_n \vdash B$ then $(A_1 \bullet A_2 \bullet \ldots \bullet A_n) \dashv 3 B$ is a theorem.

The weakness of the deduction theorem for S4 and S5 lies in that we cannot prove (2) unconditionally for $n > 1$. Whether the absence of (2) makes S4 or S5 inadequate for a systematization of the concept of deducibility, as Rosenbloom contends, would depend on a more detailed presystematic analysis of the concept. We hope to discuss this elsewhere at a later time.

8. A theorem corresponding to III for S5 was established by R. Carnap in "Modalities and Quantification," *Journal of Symbolic Logic*, XLVI (1946): 56.

9. See J. C. C. McKinsey and Alfred Tarski, "Some Theorems about the Sentential Calculi of Lewis and Heyting," *Journal of Symbolic Logic*, XIII (1948): 5. [Addendum, 1991. The Lewis systems as originally given did not include necessitation among their initial rules.]

5. *Quantification and Ontology*

This paper was originally presented at a symposium with Leonard Linsky, on Language, Logic, and Ontology, sponsored by the American Philosophical Association in May of 1971. It was published along with Linsky's paper "Two Concepts of Quantification" in *Noûs*, VI, 3 (1972). The present printing contains editorial corrections and a few clarificatory insertions. The derivation of the supposed paradox generated by substitutional semantics for sentence names has been moved from a footnote to the body of the text in order to improve the exposition.

The reader may be interested in subsequent and more detailed accounts of the substitutional quantifier by H. Leblanc, *Truth, Syntax and Modality* (Amsterdam: North-Holland, 1973); S. Kripke, "Is There a Problem about Substitutional Quantification?" in *Truth and Meaning*, ed. G. Evans and J. McDowell (New York and Oxford: Oxford University Press, 1976); Charles Parsons, *Mathematics in Philosophy* (Ithaca: Cornell University Press, 1983). ■

Interpretations of standard first-order predicate logic fail as vehicles for paraphrase of important segments of discourse. An ideally adequate interpretation of a formal language would be given in such a way that truth as well as meanings are preserved. At the same time, surface grammar would go into logical grammar; informally valid arguments would go into formally valid arguments within a language for which consistency and completeness are demonstrable. Conversely (this converse is often neglected), meaning would be preserved in translating *out* of the artificial language back into those segments of discourse for which it serves as a formalized equivalent. That such a grand program is not realizable does not and should not prevent us from enlarging our artificial languages so that we may deal with larger and larger fragments of ordinary discourse.

We are familiar with the ways in which interpretations of standard first-order logic may be used in the formalization of mathematics. But there we are dealing with areas of discourse that are already well defined and systematized. By contrast, the difficulties we encounter in attempts to accommodate discourse involving tenses, modalities, and other adverbial uses, strong conditionals such as causal implication and entailment, and the like are recalcitrant.

With a view to locating some sources of the failure of standard first-order logic (SFL), let us summarize what counts as a standard interpretation. We specify a well-defined domain of objects over which the variables are said to range. The interpretation assigns objects from that domain to the individual constants. This is done by associating with those constants names or descriptions of the objects. Predicate constants are assigned properties that in a wholly extensional language may be identified with sets the members of which have that property, i.e., its satisfaction class. This is done by associating with that predicate constant a predicate name or description. Sentence constants are assigned propositions that in a wholly extensional language may be identified with one of two truth values. Truth is a property of closed sentences. The truth of an atomic sentence is defined in terms of satisfaction of corresponding open formulae. What remains to complete the interpretation is a truth definition for the logical constants. Overly simplified, a universally quantified sentence is true iff the open sentence that follows the quantifier is satisfied by all the objects of the domain, and an existentially quantified sentence is true if the open sentence is satisfied by at least one object.[1] Truth conditions for the

---

1. Where the quantifier is followed by a closed sentence, the truth of the whole is the same as that of the sentence which follows. For a discussion of the essential role of

sentence connectives can be given in the usual way independently of the notion of satisfaction. Given in the usual way, we have an interpretation of SFL.

The failure of interpretations of SFL for translation may to some extent be a function of the *way* we specify our interpretation. Suppose we want to assign nine to a constant '*n*'. It will make a difference if the expression we associate with '*n*' is 'the number of the planets', for then the English sentence 'The number of the planets is nine', where the 'is' is taken as identity, would go into 'the number of the planets = the number of the planets' and, although truth is preserved, meaning is lost. But we could of course have associated the numeral '9' with '*n*' and restricted the terms that go with individual constants to proper *names* rather than descriptions. For singular descriptions we might adopt the theory of descriptions, or we might interpret them as unit classes or unit properties. Such modifications increase the range of translatability into and out of an interpretation, with increased retention of meaning and without tampering with the way we specify truth conditions for the logical constants.

There are more elaborate ways of giving an interpretation of SFL that, although they too do not alter truth definitions for *logical* constants, are regarded as nonstandard in the sense of going beyond some bound of normalcy for assignments to nonlogical constants on some not easily specified criterion of what are appropriate extensions of the domain. The Fregean way is such a way. There, one and the same nonlogical constant is allowed to be systematically ambiguous. It may have a multiplicity of assignments of *objects*. It may be conceived as denoting a family of objects, all of them or all but one of them an intensional object, each of which may be called by the same name. The correct reference of that name on any occasion of use is to be determined by the context. Put off by such an indefinitely expandable domain, we look for something simpler—perhaps a shift of interpretation of the *logical* constants. What I want to consider in this paper is the shift in the truth specification for quantification that I proposed in earlier papers and that has come to be known as substitutional quantification.[2]

---

a theory of satisfaction for classical quantificational semantics, see John Wallace, "On the Frame of Reference," *Synthese,* XXII (1970): 128–135.

2. See my "Interpreting Quantification" *Inquiry* (1962): 252–259; "Modalities and Intensional Language," this volume.

I

What are we to make of shibboleths like "To be is to be the value of a bound variable" or "Existence is what existential quantification expresses"?[3] They are grounded in something like this: Any consistent set of statements and its logical consequences may be thought of as a theory, and, if those statements are true, it is a true theory. If SFL is paradigmatic, i.e., if it is *the* language within which theories are to be expressed, then, since surface grammar will be replaced by logical grammar, the objects that the variables range over will be revealed to us, and those will be the very objects over which we quantify.

Now, as we know, where the subject matter is well defined, i.e., where the domain is well defined (and that is most important) and where we are *already* ontologically committed in some sense, then, all right: to be is to be the value of a variable. If we already believe— in some sense of "existence"—in the existence of physical objects or of numbers, then, if in our interpretation physical objects or numbers turn up as objects over which variables range, this squares with the status they have *already* been granted. But suppose we take the following sentences as true—"John is always late to class"; "There was at least one woman who survived the sinking of the Lusitania"; "A statue of Venus is in the Louvre"; "Necessarily, Pegasus is Pegasus, and Pegasus is not a fish"; "Zeus killed his father brutally"— how do we take them into SFL? Some defy adequate paraphrase altogether. From those that are paraphrasable some odd consequences follow. The temporal 'always' of the first sentence goes into a universal quantifier with variables ranging over temporal moments; so moments go into the domain of objects. Since the domain is unsorted, moments are there all of a piece with John, individual events, spatial locations, statues, and the like. What of Pegasus and Zeus and Venus? Since, on some prior ontological considerations, we don't want these showing up in our domain as they must if those sentences are taken to be *prima facie* true, something must be done. This requires us to devise extraordinary ways of paraphrasing that often fail to preserve meaning; e.g., "There is a statue of Venus in the Louvre" might be read inadequately "There is a statue called 'Venus' in the Louvre." It may also require that what seem to be obvious truths, like the one about Pegasus, be taken as false.

---

3. W. V. Quine, "Existence and Quantification," in *Ontological Relativity and Other Essays* (New York: Columbia University Press, 1969), p. 97.

II

Since the primary difficulty seems to be with the specification of quantification, how might we alter it? We might alter it in the direction of expanding ontology, not in the Fregean expansion of the single domain by generating a hierarchy of objects but by multiplying the domains themselves. This alternative may be represented by the semantics of Jaakko Hintikka and Saul Kripke, i.e., possible-world semantics (PWS). Here we depart from the standard interpretations of the quantifiers by admitting worlds that are possible relative to this one or, more generally, possible relative to one another. Worlds may vary both with respect to the objects in their respective domains and with respect to sentences that are true even when they are about objects common to the domains of alternative worlds. A sentence with a quantifier wholly outside the scope of a modal operator, for example, says something about objects of this world relative to other possible worlds. A quantifier inside the scope of a modal operator is appropriately relativized to the domains of those worlds. Quantification is world-bound. PWS is in a way a very natural extension of the standard semantics for quantification. Variables range over domains of individuals that are the values of those variables. Quantification is still defined in terms of satisfaction, but relativized to possible worlds.

An alternative way of altering the semantics of quantification abandons definition in terms of satisfaction and, consequently, disconnects the quantifiers from ontological commitment altogether. Like the sentence connectives, they are given in terms of truth alone. The rest is syntactical.

The idea is straightforward. On our substitutional interpretation, as in J. M. Dunn and N. Belnap's formulation,[4] we suppose denumerably infinite many constants that are associated with names. That assumption assures us that truth conditions for universal (U-) quantification and existential (E-) quantification are not equivalent to conditions on conjunction and disjunction, respectively. An interpretation assigns truth values to atomic sentences. Conditions for sentence connectives are as follows. A sentence with a U-quantifier followed by

---

4. In "The Substitution Interpretation of the Quantifiers," *Noûs*, II (1968): 179–185, Dunn and Belnap give the semantics for the substitution interpretation as follows: An interpretation (I^*) maps atomic sentences onto truth values. The *valuation v* determined by I^* for a sentence A is given by

1. If A is atomic, $v(A) = I^*(A)$.
2. If $A = {\sim}B$, $v(A) = \text{T}$ iff $v(B) = \text{F}$.
3. If $A = B\&C$, $v(A) = \text{T}$ iff $v(B) = \text{T}$ and $v(C) = \text{T}$.
4. If $A = (x)B(x)$, $v(A) = \text{T}$ iff $v(B(t)) = \text{T}$ for all *names t*.

an open sentence is true iff the open sentence is true on all substitutions of names for the variables bound by the quantifier. An E-quantified sentence is true iff it is true on one such substitution. We will call the results of such a replacement a "substitution instance" of the quantified sentence.

The conclusions and misapprehensions that originally surrounded this simple proposal may be traced perhaps to the fact that the standard semantics had assumed the role of a paradigm firmly entrenched in the conceptual scheme of logicians and philosophers.[5] Indeed, some[6] who were initially strong critics have slowly shifted until now criticism is of a much more elusive kind. In what follows I want to make explicit the initial ground for the proposal and also to answer some questions raised by Leonard Linsky in connection with extending the interpretation to systems where the syntactical categories for substitution classes are other than singular proper names. The latter is interesting since it further illuminates singular existence anomalies that have to do with substitutivity and identity.

III

The impetus for the initial proposal was not, as sometimes suggested, wholly grounded in trying to find a way of quantifying into and out of modal contexts. Nor was it merely a nominalist disposition that sought ontological neutrality at any cost. It was rather the much more general observation that there is a genuine question about the appropriateness or even the meaningfulness of supposing that there is an unequivocal connection between the standard interpretation of the quantifiers and any paraphrase into and out of ordinary and philosophical discourse. The standard semantics *demands* a clearly specifiable domain over which the variables range and which comprises their values. The description of that domain must be sufficiently clear so that we can distinguish objects that are in it. In the formalization of arithmetic we can characterize the domain by describing a structure,

5. Dunn and Belnap list these misapprehensions and answer to them, ibid., pp. 184–185.

6. See, for example, W. V. Quine, "Reply to Professor Marcus," in *The Ways of Paradox* (New York: Random House, 1966), pp. 175–182; "Ontological Relativity" and "Existence and Quantification," in *Ontological Relativity and Other Essays,* pp. 26–68, 91–113. See also J. Wallace, "On the Frame of Reference," *Synthese,* XXII (1970): 128–135. Also H. Leblanc, "A Simplified Account of Validity and Implication for Quantificational Logic," *Journal of Symbolic Logic,* XXXIII (1968): 231–235.

the objects of which satisfy the axioms of arithmetic. One might perhaps be able to arrive at such a characterization for a domain of physical objects, although attempts toward this end require specification of identity conditions for physical objects that are far from apparent. Then what, if we are dealing with ordinary and philosophical discourse, is the clearly specifiable domain over which the variables range? What is the role of the variables? There is a hint here and there in some pronominal usage.

R. M. Martin, in answer to questions like those raised above, says:

> Very well then Quine might answer, take *D* (the domain) as consisting of *all objects*. The question then arises, what meaning can be given to the phrase ''all objects''?

> Quine of course prefers physical objects to others. . . . set theorists demand another domain . . . but the theologian needs another and the literary critic still another and so on. . . .

> . . . Nothing whatsoever is gained by insisting on a reading ''for all objects *x*''. In fact quite the contrary. What are we to include as objects? Angels, dreams, dispositions, works of art, values etc., Russell's class of all classes which are not members of themselves? Sentences or inscriptions which say of themselves that they are not true?[7]

Martin seems to conclude that standard semantics requires that we retreat from the uses to which logic is put in the analysis and paraphrase of ordinary and philosophical discourse. Substitutional quantification suggests a retreat instead from the received semantics, or at the least holding it in abeyance. Ground the semantics of quantifiers in the notion of truth, but beyond that take it as a syntactical matter requiring no further specification of domains and no notion of satisfaction. I agree with Martin in this: the quantifiers cannot always be read 'There exists an object . . .' or 'For all objects . . .'. But the quantifier can be used as a logical constant, an operator required for characterizing validity in a more neutral way. We can then explore without regimentation the conditions under which quantification does or does not have existential import, for clearly there is an important connection. With the substitutional interpretation analyticity and validity *can* be characterized. The *logical* uses of quantification are preserved. The occasions on which a true sentence beginning with an E-

7. R. M. Martin, ''Existential Quantification and the 'Regimentation' of Ordinary Language,'' *Mind*, LXXI (1962): 527.

quantifier *can* be read "there exists . . ." are not discovered by logic alone. E-quantification is ontologically neutral. The matter may be put in this way: The standard semantics *inflates* the meanings of sentences it paraphrases, those, for example, that did not *originally* have the existential import they acquire on such paraphrase. When ontological inflation is avoided, the apparent anomalies that arise in going from so simple a sentence as

A statue of Venus is in the Louvre.

to

$(\exists x)$ (A statue of x is in the Louvre).

are dispelled. Whatever the *ontological* status of Venus, it is not something conferred by the operation of E-quantification, substitutionally conceived.

Given the freedom of logical operators from ontological bias, anomalies of quantifying in and out of a wide range of intensional contexts are similarly dispelled, and, as we will note below, there are also interesting consequences for identity and substitutivity.

The substitutional interpretation should also have special appeal for those of nominalist disposition who have been shy of extensions of SFL to other ontological categories, on the belief that such extensions committed them to the existence of entities like propositions or properties. Quantification does not by itself confer existence, and nominalists need not demur.

But it is a misapprehension to suppose that I see in substitutional quantification a complete account of generality. The substitutional interpretation yields a minimal account. It frees us to explore generality and existential import in all their subtlety. The satisfaction theory, by contrast, is a kind of logician's monism. But it does not seem to me that the existence of sets or numbers or propositions or attributes or physical objects hinges wholly on the way variables and quantification function in theories. Existence hinges rather on criteria of evidence and presuppositions about identity.

IV

I should like now to consider some questions raised by Leonard Linsky in connection with the substitutional interpretation where it is extended to other syntactical categories. In my original discussions[8] singular proper names were taken as the substitution class. One can choose alternative substitution classes, for example, sentences. Quotation contexts may be seen as a limiting case of obliquity, or what has been called, on the objectual view, "referential opacity." On the substitutional view, quantification into and out of quotation contexts, like quantification into and out of the epistemic contexts, makes sense. Michael Dunn and Nuel Belnap[9] point to such extensions as of particular interest in conjunction with the formalization of syntax. However, if we do take sentences as our substitution class and introduce quotation contexts, we can then define a truth predicate. And that opens the door to paradox.[10] An abbreviated version of Linsky's derivation of the paradox is as follows:

Truth definition:

(α) $Tx =_{\text{df}} \sim(p)((x = \text{`}p\text{'}) \supset \sim p)$.

Let 'S' abbreviate

$(p)((c = \text{`}p\text{'}) \supset \sim p)$.

Let 'c' abbreviate

the sentence on this page that is below 'Let 'S' abbreviate'.

The following are taken as true. The first (β) is taken as an obvious assumption, the second (γ) follows from the above abbreviations.

(β) $(p)(q)((\text{`}p\text{'} = \text{`}q\text{'}) \supset (p \equiv q))$

(γ) $c = \text{`}S\text{'}$

Proof:

(1)	$(p)((c = \text{`}p\text{'}) \supset \sim p)$	Premise
(2)	$(c = \text{`}S\text{'}) \supset \sim S$	Univ. Instant.
(3)	$S \supset \sim S$	γ, Mod. Pon., Cond.

8. "Interpreting Quantification"; "Modalities and Intensional Language," this volume.

9. "The Substitution Interpretation of the Quantifiers," p. 185.

10. L. Linsky, "Two Concepts of Quantification," *Noûs*, VI (1972): 224–239. [Linsky's paper is partially included in the appendix of *Names and Descriptions* (Chicago: University of Chicago Press, 1977); he added what is shown here, i.e., that there is no paradox if conditions of definitional adequacy are met. Addendum 1991.]

(1)	$\sim((p)(c='p')\supset\sim p)$	Premise
(2)	$(c='p')\&p$	Exist. Instant.
(3)	$'S'='p'$	γ, Simpl., Ident.
(4)	$\sim S\supset S$	β, Simpl., Mod.Pon., Cond.

It is to be expected that unrestricted higher-order extensions of first-order logic will, as they have heretofore, generate antinomies, independent of the particular case we are here considering. In the present case, no *post hoc* or seemingly artificial restrictions are requisite, for what we see is that the semantics (i.e., the truth definition α) for the quantifiers does not meet minimal criteria for definition when it is extended to the substitution class of sentences.[11] On the substitution interpretation, the truth condition for a sentence beginning with a universal quantifier for the substitution class of names is given, following Dunn and Belnap,[12] as

(Q) If $A=(x)B(x)$ then $v(A)=\mathrm{T}$ iff $v(B(t))=\mathrm{T}$ for all *names t*.

If we restrict our names to proper names and adopt for singular descriptions a theory of descriptions or a suitable alternative, no additional conditions on (Q) are required. The definiens reduces the complexity of the definiendum; i.e., the number of quantifiers is reduced, and the definition is recursive. If, as in Dunn and Belnap, we include descriptive terms in the class of names, then a condition on substitution of names analogous to that discussed below for sentences, will be required to ensure the adequacy of the definition.

Consideration of definitional adequacy requires that, for the substitution class of *sentences,* the condition on quantification be

(Q′) If $A=(x)B(x)$ then $v(A)=\mathrm{T}$ iff $v(B(t))=\mathrm{T}$ for each *sentence t* such that $(B(t))$ contains fewer quantifiers than A.

Since (Q′) places restrictions on the rule of Universal Instantiation, the first limb of Linsky's proof would fail. He cannot derive his '$\sim S$', i.e.,

$$\sim(p)((c='p')\supset\sim p)$$

But, if we grant him his other assumptions, he can derive 'S', and from his definition of truth, it follows that

c is not a true sentence.

11. The inadequacy of the truth condition on quantifiers where extended to other syntactical categories was also pointed out by Bas van Fraassen, in a letter that was circulated in, I believe, 1969. (The letter is undated.) My proposed modification is not patchwork. Definitions must reduce complexity.

12. "The Substitution Interpretation of the Quantifiers."

which is as it should be, for the sentence '*S*' that says of itself that it is not true, is true.

V

Although the substitution interpretation doesn't force extensions to other syntactical categories, it makes such extensions plausible, since it frees us of the ontological baggage of objectual quantification. Furthermore, it gives coherence to the claim that quotation contexts may be seen as strong obliquity. I would like to consider this last claim, for it helps us to articulate the way in which substitutional quantification permits a rational rethinking of problems connected with identity and substitutivity.

Identity is generally understood to be the strongest equivalence relation that an object has to itself. Furthermore, we are told, "Substitutivity is a fundamental principle governing identity."

Now, as we know, substitutivity does not in fact govern identity. We are familiar with the counterexamples, and those counterexamples constitute a recalcitrant problem for the standard interpretation. Given the objectual view, individual symbols are assigned *objects,* and what is true of an object should remain true, whatever its name. The Fregean solution assigns a multiplicity of related objects of increasing intensionality to individual names and descriptions. Possible-world semantics multiplies domains instead, but makes univocal assignments in each possible world. Substitutional quantification is not grounded in assignments of objects at all. It can therefore introduce a logical constant, call it 'I', which is in fact *defined* by the principle of substitutivity and therefore governed by it in the strongest sense. We are then free to explore what precisely the connection is between substitutivity and identity.

The semantics of the substitutional interpretation for first-order logic may be extended as follows:

(I)   If $A = s\mathrm{I}t$, $v(A) = \mathrm{T}$ iff for all sentences B_1 and B_2 where B_2 is like B_1 except for containing occurrences of the name t at one or more places where B_1 contains the name s, $v(B_1) = \mathrm{T}$ iff $v(B_2) = \mathrm{T}$.

Let us assume that the *substitution* class of names does not include quotation names, although sentences may contain quotation names. This is consistent with taking a quotation context as an oblique context with quantification in and out but not over. Now we can, on the substitutional view, be free with respect to the standardized predicates we associate with predicate constants. They can be any predicates formed from sentences by replacing names from the substitution class of names by place markers. Given such unrestricted formulation of predicates, it is clear that, if we allow quotation-name contexts, then, where s and t are quotation names,

$$sIt = F$$

except where s and t are the same name, i.e., s and t are tokens of the same type or equiform. For, although

> Tully is the same as Cicero.

is true,

> 'Tully' is the same as 'Cicero'.

is false.

Let us suppose that in *epistemic* contexts like

> John doesn't know that Cicero is Tully.

John is not being said to be ignorant about whether something is the same as itself. The statement above might therefore be represented as

> John doesn't know that Cicero is called 'Tully'.

or

> John doesn't know that 'Tully' names Cicero.

or

> John doesn't know that 'Tully' and 'Cicero'' name the same thing [person].

There are other alternatives. An analysis would depend on the context. These too are quotation contexts, but not so *all* epistemic contexts. Now, if we exclude from our standard predicates those that contain place markers inside quotes, then our logical constant 'I' may be taken as definitive of synonymy for names. What we note is that 'I' is an equivalence relation that is relative to the permissible contextual obliquity on a given interpretation of our formal language.

This suggests *where,* finally, the logical constant 'I' connects with ontological problems of identity. Leaving aside for the moment the question whether it makes sense to talk of absolute identity, there is the familiar problem of specifying minimally when an x and a y are the same of a kind, i.e., the same person, the same pain, the same physical object.

Consider the case of physical objects. Suppose there is some clear criterion for marking out a property as a property of physical objects *qua* physical object. Suppose we associate with our predicate constants, standardized predicates corresponding to those properties. Predicates that include place markers in oblique occurrences other than modal would be largely excluded[13] for the reason that, for example, beliefs about physical objects would not count among the properties of those objects *qua* physical object. And suppose we restrict the predicates to just such predicates. On such an interpretation, substitutionally conceived, our contexts are physical-property contexts. Therefore, if

$$sIt = T$$

then the names *s* and *t* are intersubstitutable in all physical-property contexts. But there is a further link missing. It is generally held by philosophers that, in a true statement of physical identity, the terms must denote an actual physical object. For the given interpretation we should then add that if, furthermore, *s* and *t* denote, then they denote one and the same physical object; the "objects" they denote are identical physical objects objectually understood.

13. I hedge here in saying "largely excluded." If contexts involving such locutions as "observes that" are construed as oblique, then we cannot presume that all such predicates can be excluded.

6. *Classes, Collections, Assortments, and Individuals*

In August of 1962 a colloquium was held in Helsinki the papers of which were published in *Acta Philosophica Fennica,* XVI (1963). It was a memorable occasion. I especially recall Jaakko Hintikka's "The Modes of Modality," Saul Kripke's "Semantical Considerations on Modal Logic," Richard Montague's "Syntactical Treatments of Modality," Peter Geach's "Quantification Theory and the Problem of Identifying Objects of Reference," and Arthur Prior's "Is the Concept of Referential Opacity Really Necessary?" An extraordinary feast of reason and imagination. My contribution was "Classes and Attributes in Extended Modal Systems." The present paper, submitted to *American Philosophical Quarterly* in 1965 under the title "Classes, Collections, and Individuals" is an abbreviated version of the colloquium paper in which I had *inter alia* a struggle with arriving at a salient vocabulary for the distinctions I wanted to make among notions that are often conflated in varying ways, such as attribute, class, collection, set, and what I have in the present paper called "assortment." I arrived finally at a terminology that is at variance with that of my initial effort, but the distinctions to be captured remain the same. I continued to worry about the best way to make my exposition clearer and asked that publication be deferred. In the end, the present paper was published in 1974 as submitted under the title "Classes, Collections, and Individuals," when APQ's patient editor presented me with an ultimatum.

For a more recent development of similar views, see Charles Parsons, *Mathematics in Philosophy* (Ithaca: Cornell University Press, 1983), especially chapter 11, "Sets and Modality."

The present printing includes some editorial changes, corrections, and a few clarificatory insertions. ∎

The logical theory of sets obscures commonplace distinctions that we make between assortments, collections, and classes, in particular, in those commonplace uses where we speak of collections of physical objects. In this paper I shall suggest that the puzzles generated by assimilating ordinary collections to classes reflect the same difficulties as those puzzles that follow from equating proper names with definite descriptions. An indication will also be given as to how, within an alternative logical framework, these distinctions can be articulated.

Consider the following example from an elementary mathematics text:

> In a club with five members, there are 6 possible committees with one member, 10 committees with two members, 10 committees with three members, 5 committees with four members, 1 committee with five members. The selection of a committee which includes all five members (sometimes referred to as the "committee of the whole") means that there are zero members not serving. Thus we may balance our table by saying that there is 1 possible committee with zero members. (School Mathematics Study Group, *Mathematics for High School* [New Haven: Yale University Press, 1960], vol. II, pt. II, p. 284)

To take the above quotation seriously is to be puzzled. Clubs as we know them have constitutions and bylaws. They usually define each of their committees not by naming particular individuals as members but by specifying its size and function and how its members are to be selected (e.g., "The Executive Committee shall consist of the officers and one person elected by a majority vote of the membership. Its duties shall be. . . .") Suppose *a, b, c, d, e,* are the five members of the club described, and suppose the same three persons *(a, b, c)* were elected to the Nominating Committee and to the Resolutions Committee. (Each a committee of three.) According to the text quoted, they are the same committee. But are they? Would the Resolutions Committee as Resolutions Committee propose nominations? Would the Nominating Committee as Nominating Committee propose resolutions? Isn't it merely a historical contingency that the same persons serve on both committees, just as it is a historical contingency that *a, b, c* compose the Resolutions Committee? It might have been *c, d, e.* Similarly for the Nominating Committee. Yet, it is certainly true, given the accomplished fact, that every member of the one is a member of the other. There is then a kind of accidental equivalence between those committees. But surely it is not a historical accident that the collection of persons *a, b, c* is the same as the collection of persons *a, b, c.* Indeed, if the bylaws were such that each of these two committees was a committee of one rather than three, to which the same member had been elected, we would discern shades of the evening star and the morning star.

The committees of the example correspond to classes where composition is fixed by describing conditions of membership. But the individuals *a, b, c* compose a collection or aggregate of specific individuals. The theory of classes (or sets), by virtue of some of its fundamental assumptions, conceals such differences. Yet 'class' and 'collection' in logic are not held to be mere homonyms of colloquial counterparts. In naïve and often in axiomatic set theory, they are introduced in terms of correlative informal notions. Departures from familiar meanings are taken to be technical extensions, modifications, or refinements. Consider the following examples:

(1)   The essential point of Cantor's concept is that a collection of objects is to be regarded as a single entity (to be conceived as a whole). The transfer of attention from individual objects to collections of individual objects is evidenced by the presence in our language of such words as 'bunch', 'covey', 'pride', and 'flock'.

        With regard to the objects which may be allowed in a set (Cantor's description), "objects of our intuition or of our intellect" gives considerable freedom. . . . Green apples, grains of sand or prime numbers are admissible constituents of sets. (R. R. Stoll, *Introduction to Set Theory and Its Logic* [San Francisco: Freeman, 1961], pp. 2–3)

(2)   A pack of wolves, a bunch of grapes, or a flock of pigeons are all examples of sets of things. . . . To avoid terminological monotony, we shall sometimes say *collection* instead of *set*. The word 'class' is sometimes used in this context. (P. Halmos, *Naïve Set Theory* [New York: Van Nostrand, 1960], p. 1)

(3)   By a *set* we mean any kind of collection of entities of any sort. . . . Many other words are used synonymously with 'set': for instance, 'class', 'collection', and 'aggregate'. (P. Suppes, *Introduction to Logic* [New York: Van Nostrand, 1959], p. 177)

The common definition of "collection" as

(4)   A group of things that have been gathered together.

seems adequate to familiar examples such as coin or art collections, or the sundry objects in a woman's purse. Yet there seems to be an inescapable peculiarity in the shift (as in (1)) to entities such as numbers, as if they could be gathered together like pebbles on a beach. Is it merely an extension of a familiar notion to include objects of our intellect as members of collections in accordance with Cantor's proposal? Medals may be collected, but what of virtues? The response to such naïve questioning is that no such literal gathering together is intended.

(5)   We can say that a class is any aggregate, any collection, any combination of objects of any sort; if this helps well and good. But even this will be less help than hindrance unless we keep clearly in mind that the aggregating or collecting or combining here is to connote no actual displacement of objects, and further that the aggregation of say seven given pairs of shoes is not to be identified with the aggregation or collection or combination of those fourteen shoes nor with that of the twenty-eight soles and uppers. In short, a class may be thought of as an aggregate or collection or combination of objects just so long as 'aggregate' or 'collection' or 'combination' is understood strictly in the sense of 'class'. (W. V. Quine, *Set Theory and Its Logic* [Cambridge: Harvard University Press, 1963])

But to remain unabashed by the circularity is not to resolve the perplexity. If abstract objects such as numbers are eligible to be elements of collections, does it make sense to talk of displacement at all? And what principle precludes the identification of an aggregate of seven pairs of shoes with those fourteen shoes? It would be difficult to persuade a shopkeeper.

Implicit in the ordinary use of 'collection' (4) is that (a) collections are not empty, and (b) they consist of distinct and distinguishable objects. Also, since 'gathered together' and 'aggregated' are in the past tense, (c) collections are in some sense complete. But what kinds of objects are they, how are they distinguished, and in what sense have they been "gathered together"? Confusion is minimized if for present exposition, we restrict the entities of a collection to individuals that are distinguishable as physical objects. And indeed this is the colloquial use of the term. Since they have been gathered together, the number of objects is finite, whether they be assembled in heaps or piles, or hung on museum walls. Nevertheless, even here 'gathered together' is difficult to make precise. Generally, spatiotemporal prox-

imity is involved, though, as in the case of an art collection, not necessarily. The common feature of such aggregates is that they consist
of enumerably many individual physical objects. How many? Presumably some finite plurality depending on the particular mode of
aggregation. How many grains of sand make a heap, or grapes a
bunch?

We do not have a common word for

(6)   An *arbitrary* finite selection of physical objects

that is entirely separable from connotations about the *mode* of aggregation. In the absence of a more felicitous alternative, I shall use
'assortment' for (6). I do not propose here to go into an analysis of
the notion of physical object or to raise questions as to whether, for
example, shadows or afterimages qualify. I shall assume that the predicate is sufficiently well understood so that particular physical objects
can be individuated.

It is clear from the above that logical classes are very different
from assortments. They may consist of any entities whatever, including entities that may themselves be classes. Logical classes may
be empty or infinite. The number of members of a logical class, even
where the members are taken to be physical objects, may be indefinite
in a historical sense, as when we speak of the class of animals or stars,
since entities are still being born.

An exhaustive account is given of an assortment by specifying each
of its members. But the extended notion of class precludes such an
account in almost all cases. If the common definition of collection,
as in (4), is to apply to logical classes as well, it must be in some
other sense of "gather," as when Kant speaks of "gathering together
under a concept" or "gathering together under a rule." The theory
of classes or sets obscures the distinction between ways of "gathering
together," or, using Cantor's phrase, "modes of conceiving a totality."

What I have suggested above is that there is an ambiguity in the
notion of "gathering together" and "conceiving a totality" that is
reflected in the distinction between an itemization of elements and a
statement of conditions for membership. The obliteration of the difference by the logic of set theory does not generate too many perplexities for mathematics (leaving aside for the moment paradoxes of
the Russell and Skolem sort) since the elements with which mathematics deals are *defined* by those conditions, i.e., the rules or concepts
under which the elements are "gathered" together. A number cannot
fail to satisfy any one of its arithmetically unique conditions and still

be that number. It makes no sense to say that 3 is no longer the whole number between 2 and 4 but is still 3. But it does make sense to say that Venus has changed its orbit but is still Venus. An entity such as a number may therefore be equated with any one of its mathematically unique descriptions without generating many of the familiar puzzles that follow upon equating the proper name of an object such as a physical object with a definite description—other than a definite description that specifies as the unique condition that it *is* that object, e.g., $\imath x(x\mathrm{I}a)$ where 'I' stands for the identity relation between individuals and '*a*' is the individual object's proper name.

The absence of differentiation between these two modes of conceiving a totality is apparent from the manner in which sets are described. For example, one reads in Patrick Suppes's[1] *Introduction to Logic:*

(7)   Often we shall describe a set by writing down the names of its members, separated by commas and enclosing the whole in braces. For instance by

(a)   {Roosevelt, Parker}

we *mean*

(b)   the set consisting of the two major candidates in the 1904 American presidential election.

By

(c)   {1, 3, 5}

we *mean*

(d)   the set consisting of the first three odd positive integers.

In (7a, c) we find an exhaustive list of names that is bracketed, whereas in (7b, d) what we have is a specification of conditions to be met. Yet (7a) is taken here to have the same *meaning* as (7b), and similarly for (7c) and (7d). No great difficulties are encountered in connection with the numerical case if we do not take a numeral to have the referential function of a proper name. But here, too, if the numerals *are* taken to be proper names in a strict sense and if we

1. New York: Van Nostrand, 1959, p. 179. ['Class' and 'set' are in Suppes taken as synonyms. Addendum, 1991.]

attempt to track down the objects of reference, difficulties develop.[2] It would be much like trying to find *the* object of reference of the white queen of chess. Any *candidate* whatever for white queenship must move in accordance with the rules, and beyond that there would be no way of deciding between rivals for the job.

The more familiar difficulties are connected with equating (7a) and (7b); taking as synonymous a designation of an assortment that, as suggested above, consists of an inventory, and a description that fixes a range of objects, not by an inventory but by specifying the conditions to be met by undesignated objects. A well-known example will serve to remind us of these puzzles.

According to (7)

(8) (a) $\{N, M, V, P, MA, U, E, S, J\}$ (where 'N', 'M', . . . abbreviate the proper names of the planets)

means the same as

(b) the class of planets

But surely not. Venus's membership in the *assortment* designated by (8a) is a matter of logical necessity, but not its membership in the *class* of planets. The number of objects in the assortment (8a) is *logically* fixed; not so the number of planets. We may know that Venus is contained in an assortment that has Venus among its elements, and not know that it is a planet, and so on.

But (7) should not be taken as a truth claim about synonyms that have been discovered to hold between expressions designating assortments by inventory and expressions that describe classes without specification of individual members (although it is sometimes mistaken as such a claim). Rather, proposals such as (7) should be understood as decisions to ignore the difference between what we have called assortments, and classes. By virtue of such a decision, we arbitrarily restrict talk of classes to truth-functional extensional contexts. If we adhere to this decision, we would, for example—allowing for the possibility that by some cosmic cataclysm Venus might be hurled out of its orbit around the sun—have no way of asserting that it is possible that Venus not be a planet. Such an assertion would generate a contradiction. The systematic manifestation of principles such as (7) is, of course, the postulation of a principle of extensionality: classes that have the same members are the same class. There is an obviousness

---

2. See for example, the interesting paper of Paul Benacerraf, "What Numbers Could Not Be," *Philosophical Review*, LXXIV (1965): 47–73.

about such a principle, but only because the intuitive notion of a class is what we have here termed an "assortment." And, indeed, for assortments, extensionality need not be postulated at all. But classes are not conceived of as an actuality of finitely numerable objects. Rather, they correspond to attributes and are generated by a general principle of set formation. Such an *Aussonderungsaxiom* states roughly that

> (9)   To every *wff A* that contains *a* as the only free variable, there corresponds a class *r,* and to every class *r* there corresponds a *wff A* such that $A = a \in r$.

If the device of abstraction is used, *r* is the abstract of *A*. The symbol '=' may, as in *Principia Mathematica,* signify equality by definition, in which case the abstraction operator '∧' and the connective for membership '∈' are taken as part of the contextual definition of classes. Where (9) is introduced axiomatically the abstraction operator is generally taken as primitive, and '=' is some systematic relation of equivalence such as material equality, or a stronger equivalence if it is available as in some of the modal systems.

It is well known that (9) is not acceptable unqualifiedly, in view of Russell's paradoxes and perplexities stemming from the Löwenheim-Skolem theorem. Our choice of a solution is type-theoretic, not merely because it resolves the Russellian paradoxes but also because the notion of an assortment is grounded in the supposition that there are at least individual *physical* objects that are not themselves sets or classes.

Consider such a type-theoretic applied predicate calculus of second order with individual and predicate variables and constants, and abstraction. Individual constants are interpreted as proper names of physical objects. Under this interpretation, identity is a relationship that holds between an individual and itself. As such, an identity claim is not a contingent matter. We are supposing here that there is a purely referential function of expressions that is the primary use of proper names.[3] A proper name does not have a predicative function, whereas a singular description predicates uniquely. Although self-identity is not a contingent matter, it may be a contingent matter that two singular descriptions apply to the same individual. This suggests that there are weaker equivalence relations proper to descriptions. One systematic representation of these differences is roughly to construe singular de-

---

3. See "Modalities and Intensional Languages," this volume, for a discussion of the purely referential role of proper names and the principle of the necessity of identity.

scriptions as (1) unary abstracts on the level of first-order predicates, which (2) contain a uniqueness condition.

The discussion of proper names is not merely tangential to the consideration of classes and assortments, for the analogies are close. If, to extend a metaphor of John Searle's,[4] proper names are pegs on which to hang (singular) descriptions, then a bracketed list of proper names (which designates an assortment) is a peg on which to hang a class description, as in (8a, b). Just as "the immense pragmatic convenience of proper names in our language lies precisely in the fact that they enable us to refer publicly to objects without being forced to raise issues and come to agreement on what descriptive characteristics exactly constitute the identity of the object," so we may, through a device like bracketed lists of proper names, refer publicly to groups of objects, without being committed to any one identifying descriptive characterization of the assortment other than that which characterizes it *as* an assortment. As in the case of proper names, an equivalence relation between assortments is never contingent, but an equivalence relation between classes may be.

If we are to be able to represent such distinctions between various equivalence relations, such as identity of individuals and logical, material, and contingent equality of classes, within the object language of the predicate calculus, we shall need in addition to the truth-functional connectives a stronger implication.

In a previous paper[5] I discussed the formal details of such an intensional modal system ($S4^2$) that is based on an extension to second order of S4. In it we may derive metatheorems that correspond to the following:

(10)   A principle of extensionality holds for assortments.

(11)   A principle of extensionality holds for classes that are strictly equal ($=$). For materially but not strictly equal classes ($=_m$), the principle of extensionality is restricted.

An elaboration of (10) and (11):

(12)   Two classes may be weakly equivalent (materially or contingently). If each of them is weakly equivalent to some

4. J. R. Searle, "Proper Names," *Mind* (1967): 166–173. On Searle's view names are mere pegs. On my view they have fixed reference.

5. "Classes and Attributes in Extended Modal Systems," *Acta Philosophica Fennica, Proceedings of a Colloquium in Modal and Many-Valued Logics* (1963): 122–135, which is an application of "A Functional Calculus of First Order Based on Strict Implication," *Journal of Symbolic Logic*, XI (1946) and "The Identity of Individuals in a Strict Functional Calculus," XII (1947). The Barcan formula and its converse are theorems.

assortment, the corresponding assortments are the same (strictly equal).

A word of elaboration about (10–12).[6] It should be noted that I said "A principle of extensionality" rather than "The principle of extensionality." The strongest claim would be that if classes have the same members, then they are identical. In S4^2 the identity relation is defined not over classes but over individuals. However, the systematic manifestation of an extensionality principle is an unrestricted substitution rule. Such a substitution rule is provable for assortments and for classes that are *strictly* equal. Our interpretation of S4^2 would still exclude oblique epistemic contexts, since, although two classes may be logically equal, the alternative abstract-descriptions may not, it seems, be intersubstituted in such contexts without going from truths to falsehoods. Consider, for example, the strictly equal $\hat{x}(x\text{ITully})$ and $\hat{x}(x\text{ICicero})$, where substitution might fail in epistemic contexts.

Returning now to S4^2, the abstraction operator ($\wedge$) is taken as primitive, and an *Aussonderungsaxiom* is postulated. For unary abstracts with no free variables, it takes the form (omitting precautions about choice of variables)

(13) $(b\in\hat{a}A) \equiv B$, where a is the only individual variable occurring freely in A, and B results from replacing all free occurrences of a in A by b

Here and in what follows, '$\equiv$', '$\equiv$', '$\dashv$' and '$\supset$' are symbols for strict and material equivalence, strict and material implication respectively. Substituends of individual variables are nondescriptive individual constants, i.e., proper names. The notation r, s, t will be used for unary abstracts. They may be understood as designating classes.

Definitions of sum, product, complement are introduced in the usual way. The two implication relations ($\supset$, $\dashv$) and the two equivalence relations ($\equiv$, $\equiv$) have their Boolean analogues: material and strict inclusion ($\subset$, $\in$), and material and strict equality ($=_m$, $=$). We shall have reason to refer to the latter definitions, which may be given as

(14) $(r=_m s) =_{df} (\forall a)((a\in r) \equiv (a\in s))$

---

6. A somewhat different vocabulary is used in "Classes and Attributes in Extended Modal Systems" to capture distinctions between what I here call "assortments" and "classes." Classes are here equated with attributes. Assortments may be thought of as corresponding to attributes given by inventory.

(15)　　$(r = s) \;=_{df}\; (\forall a)((a \in r) \equiv (a \in s))$
　　　　or $\;=_{df}\; \square (r =_m s))$

The universal and null terms are defined as

(16)　　$V =_{df} \hat{a}(a I a)$ and $\wedge =_{df} - V$

An assortment may be represented as follows:

(17)　　$\hat{a}A$ designates an assortment if there is a B such that A may
　　　　be transformed into B by elimination of defined constants
　　　　and by application of rules of association, distribution, and
　　　　double negation, where B is of the form

　　　　　$\hat{a}((a I b_1) \vee (a I b_2) \; . \; . \; . \; (a I b_n))$

　　　　'I' names the identity relation, and $b_1, b_2, \ldots, b_n$ are
　　　　individual constants (proper names).

Given that the necessity of identity for individuals holds in S4^2, it will
follow that membership in an assortment is necessary. It can also be
shown by an induction on the number of disjuncts that a principle of
extensionality holds for assortments *without* further assumptions.

(18)　　Let r designate an assortment as in (17) of individuals b_1,
　　　　$b_2, \ldots, b_n$ $(n \geq 1)$ and s an assortment of individuals
　　　　$c_1, c_2, \ldots, c_n$ $(n \geq 1)$. If $r =_m s$ then $r = s$.

For it is also provable that

(19)　　If $r = s$ then 'r' may replace one or more occurrences of
　　　　's'.

It should also be noted that unrestricted substitution of 'r' for 's'
is not available where

　　　　$(r =_m s)$ & $-(r = s)$

However,

(20)　　If $r =_m s$ and if 's' does not occur within the scope of a
　　　　modal operator, then 'r' may replace one or more occur-
　　　　rences of 's'.

By virtue of (20), in the case of material equality, substitution is permitted only in truth-functional contexts.

The more formal statement of (12) reads

(21)　　If $r_1 =_m r_2$, $r_1 =_m s_1$, $r_2 =_m s_2$ and if 's_1', 's_2' are abstracts
　　　　which designate assortments, then $s_1 = s_2$.

99

(21) is an analogue for assortments of the necessity of identity for individuals.

Returning now to the original example. The Nominating Committee, N, may be represented by an abstract over a set of defining conditions specified in the bylaws. Similarly for the Resolutions Committee R. The assortment (or collection or aggregate) (C) consisting of three members a, b, c may be represented as

$$(22) \quad \hat{x}[(xIa) \lor (xIb) \lor (xIc)]$$

Then if

$$(23) \quad \text{(a)} \ (R =_m N) \ \& \ \Diamond \sim (R =_m N)$$

$$\text{(b)} \ (R =_m C) \ \& \ \Diamond \sim (R =_m C)$$

$$\text{(c)} \ (N =_m C) \ \& \ \Diamond \sim (N =_m C)$$

'N', 'R', 'C', are all intersubstitutable in, for example,

$$\text{(d)} \ (a \in R)$$

but not in

$$\text{(e)} \ (R = R)$$

for, when (23e) is expanded according to (15), it can be seen that in (23e) 'R' is in the scope of a modal operator. Suppose, furthermore, that 'a_1', 'b_1', 'c_1' are alternative proper names of a, b, c, respectively. Call the assortment "D." Since it is also the case that

$$\text{(f)} \ (N =_m D)$$

it follows from (23c, f) that

$$\text{(g)} \ C =_m D$$

But 'C' and 'D' designate assortments, and from (21) it follows that

$$\text{(h)} \ C = D$$

This is not a case of converting contingencies into necessities by symbolic hocus-pocus. That a, b, c, by whatever name we call them, happen to have been elected to two committees is interpretable as a contingent fact. But it is not a contingent matter that, however we tag their members, they are a single assortment. Indeed, isn't this a suitable analysis of the principle of extensionality, that it holds for assortments but not for classes? Classes (attributes) are, in Frege's language, "unsaturated."

7. Does the Principle of Substitutivity Rest on a Mistake?

This paper appeared originally in *The Logical Enterprise* (New Haven: Yale University Press, 1975), pp. 31–38, a *Festschrift* in honor of Frederic B. Fitch edited by Alan R. Anderson, Richard M. Martin, and me—a small gesture honoring a scholar whose work still remains to be fully appreciated. ■

I

In the paper "Attribute and Class,"[1] Frederic Fitch suggests that "if entities X and Y have been identified with each other, it seems reasonable to suppose that the *names* of X and Y should be everywhere intersubstitutable where they are being used as names." The principle of substitution is here proposed as a reasonable supposition. Some philosophers are less tentative in their acceptance, as is W. V. Quine when he says: "One of the fundamental principles *governing* identity is that of substitutivity. . . . It provides that given a true statement of identity one of the two terms may be substituted for the other in any true statement and the result will be true."[2]

Others are more tentative, going so far as to claim that, however reasonable a supposition the principle of substitutivity may appear to be, it is simply false. They point, for example, to the sentences

(1)   Georgione is so called because of his size.

and

(2)   Barbarelli is so called because of his size.

Here 'Georgione' and 'Barbarelli' are names being used as names, and though 'Barbarelli is Georgione' is true, substitution converts a truth to a falsehood.

Recently Richard Cartwright,[3] who rejects the principle of substitutivity on the basis of such counterexamples, has set about to explain why the belief persists that it is a governing principle. The error, he claims, is that it has been confounded with another principle, the principle of indiscernibility (or, as he prefers to describe it, the principle of identity). The latter, however, is concerned not with expressions and substitutions but rather with the properties things have. It says

(Ind)   If $a = b$ then every property of a is a property of b.

Cartwright goes on to argue that, although (1) and (2) provide a counterexample to substitutivity, it does not follow from this that such counterexamples are also counterexamples to the principle of indis-

1. In *Philosophic Thought in France and the United States*, ed. Marvin Farber (Buffalo: University of Buffalo Publications, 1950), p. 552.

2. *From a Logical Point of View*, 2d rev. ed. (New York: Harper Torchbooks, 1961), p. 139.

3. "Identity and Substitutivity," in *Identity and Individuation*, ed. M. Munitz (New York: New York University Press, 1971), pp. 119–33.

cernibility. For (1) and (2) to count as a counterexample to the principle of indiscernibility, there must be some property that Georgione has and Barbarelli lacks. But if from (1) we form the associated predicate

(3)   *x* is so called because of his size

we find that it does not determine a property. The quasi-predicate 'being so called because of his size' does not specify a function that takes us from individuals to propositions. Other attempts to specify a property attributed to Georgione in (1) and not shared by Barbarelli are shown to be incoherent. If, on the other hand, the property attributed to Georgione is that of being called 'Georgione' because of his size, this *is* a property that Barbarelli has, and we do not have a counterexample to (Ind). Cartwright goes on to examine other cases of substitution failure, such as puzzles about nine and the number of planets, and argues that here too failures of substitutivity are not at the same time failures of (Ind).

But is the connection between the principles of identity and substitutivity a case of mistaken identity? Some philosophers have clearly distinguished them from each other[4]—but have claimed as well that they are both versions of Liebniz's law: If *a* and *b* are identical, then whatever is true of *a* is true of *b*. One version, (Ind), is perceived as being in the material mode, while substitutivity is perceived as being in the formal mode. Indeed, in formal languages the two principles are intimately connected. Substitutivity may be taken as a rule of derivation in first-order logic with identity. In the absence of quantification over properties, the principle of indiscernibility comes to the set of those valid sentences which are the associated conditionals of the rule of substitutivity for each of the predicates of the theory. Alternatively, (Ind) may be taken as definitive of identity in second-order logic, and the principle of substitutivity falls out metatheoretically. Anyone who claims that substitutivity doesn't govern identity but that indiscernibility does must wonder why this is not so in formal languages. If translation into predicate logic is supposed to display the logical form of at least some important fragments of ordinary discourse, how can substitutivity rest on a mistake?

What Cartwright has noted is not peculiar to the principle of substitutivity. Apparent counterexamples can be found in natural language for virtually all logical principles where surface grammar is untrans-

---

4. See, for example, Leonard Linsky, "Reference, Essentialism and Modality," *Journal of Philosophy*, LXVI, 20 (October 16, 1969): 687–688.

formed or the resolution of ambiguities is implicit and resolvable only by attention to the context of use. Does the truth of 'Socrates is sitting' and 'Socrates is not sitting' provide a counterexample to the principle of noncontradiction? Homonyms and the absence in ordinary speech of a rule, one name one thing, generate false sentences of the form '$x = x$'. Do these count as counterexamples to a logical law?

If we are to deny (as we do) that belief in the principle of substitutivity is a straightforward case of mistaken identity, the question of the principle's logical status remains. *Prima facie,* substitutivity is not an empirical principle about language use. Is it a regulative or normative principle? An answer here depends on how one explains, or explains away, apparent failures of the principle in ordinary speech, even when ordinary speech is supplemented by some of the apparatus of formal logic.

II

Central to an understanding of the role of a principle like that of substitutivity is the belief that there are different ways of saying the same thing and that some ways are logically preferable in that they reveal logical form. Criteria for judging a way of saying as being preferable include that it resolve ambiguities and that it dispel puzzles or antinomies. Further desiderata may be more controversial. For those of a non-Meinongian temper, reduction of ontology might be a criterion. Frege states (or perhaps overstates) the case as follows:

> We must not fail to recognize that . . . the same thought may be variously expressed. . . . It is possible for one sentence to give no more and no less information than another; and, for all the multiplicity of languages, mankind has a common stock of thoughts. If all transformation of the expression were forbidden on the plea that this would alter the content as well, logic would simply be crippled; for the task of logic can hardly be performed without trying to recognize the thought in its manifold guises.[5]

One of the tasks of logic, then, is to select from among all the sentences that express the same thought those that express it better. That there is a best way, "*the* logical form," supposes, as did Wittgenstein in the *Tractatus,* that "there is one and only one complete

5. Peter Geach and Max Black, *Translations from the Philosophical Writings of Gottlob Frege* (Oxford: Blackwell, 1952), p. 46n.

analysis of the proposition.'' That supposition remains controversial. The point of this paper is simply that a principle like substitutivity is to be tested with respect to statements in logical form.

But the pursuit of logical form is multiply complicated. Logicians, until recently preoccupied with the task of selecting from among sentences that express the same proposition the ones (one) that better (best) express it, have neglected the obvious fact of usage that behind the same sentence (type) may lie different thoughts. Russell noted the ambiguity of sentences like 'George IV wonders whether Scott is the author of *Waverley*' and proposed alternative analyses for each interpretation. Similarly, we are familiar with amphibolies like 'everyone loves someone'. But, for a host of sentences, the same sentence-form may express different propositions, and the ambiguity is not to be explained in terms of structural factors. These sentences are ambiguous because a determination of the objects being talked about on a given occasion requires that we have knowledge of the wider context in which the sentence has been uttered on each occasion of its utterance. Cartwright, for example, in noting the ambiguity of (3), takes it as obvious that (1) is unambiguous in expressing a proposition. But, in the presence of a personal pronoun, that clearly is not necessarily so. Suppose (1) had been uttered in response to 'Shorty is so called because of his size'. In such a setting (1) might express that Georgione is called 'Shorty' because of Shorty's size, that Georgione is called 'Shorty' because of his own (Georgione's) size, or that Georgione is called 'Georgione' because of his own (Georgione's) size, etc.

III

What is a correct formulation of the principle of substitutivity? Cartwright takes it to be the following:

(Sub) For all expressions α and β, $\ulcorner \alpha = \beta \urcorner$ expresses a true proposition if and only if, for all sentences S and S', if S' is like S save for containing an occurrence of β where S contains an occurrence of α, then S expresses a true proposition only if S' does also.[6]

An obvious inadequacy of (Sub), if ' = ' is to be interpreted as identity, is a failure to specify a substitution class for α and β. In the absence

6. Cartwright, ''Identity and Substitutivity,'' p. 120.

of such a specification, the right side of an instance of the biconditional (Sub) could be true and the left side simply meaningless; e.g., the sentence 'just in case = if and only if' makes no sense, even though substitution of 'just in case' for 'if and only if' may be truth-preserving. The meaningfulness of an identity sentence requires in the least that the grammatical category of the terms be that of singular terms construed broadly as including demonstratives, pronouns, names, and descriptions, but a syntactic criterion is insufficient. A further condition on a meaningful identity sentence is that the terms that flank the identity sign refer. Whether or not a term refers, however, is not a syntactic matter. There are, furthermore, different ways of referring, and not all of them are verbally realized. The mere utterance 'this is identical with this', without accompanying gestures or clues is not a meaningful identity sentence; yet, truth is preserved by substitution of 'this' for 'this'. With accompanying clues, an identity sign flanked by demonstratives or pronouns is on a given occasion of use meaningful and often true, but on different occasions the same locution and gestures may refer to different objects. For such cases, although the identity sentence may be meaningful, it is not reasonable to suppose that (Sub) is a governing principle.

The remaining candidates for the substitution class in (Sub) are definite descriptions and (proper) names, but there are grounds for eliminating the former. We are speaking here of a definite description's being used in the ordinary way, namely, to pick out an object (if any, and then only one) that satisfies the description.

A digression is warranted at this juncture. Much has been made recently of the fact that a description, even an erroneous one, may serve on occasion to refer (purely) to an object.[7] That may be pragmatically interesting, but such uses are more akin to the use of demonstratives. 'The woman over there drinking a martini' may serve on a particular occasion to tag a particular man drinking champagne, but if its use is idiosyncratic and transient, it is like a discardable tag, for it does not enter into the common language. Occasionally such ways of referring become entrenched, and a measure of such entrenchment is the conversion of a faulty description to a proper name. Ordinary English has a way of marking the change by using capitals. 'The Evening Star' serves as an example. It is curious that so simple a typographical device for separating the proper-name-like use of a description from its predicative use has never been employed in formal

---

7. K. S. Donnellan, "Reference and Definite Descriptions," *Philosophical Review*, LXXV (1966): 281–304.

analysis. How directly the apparent contradiction in 'The evening star is not a star' is dispelled by 'The Evening Star is not a star'!

Unlike demonstratives and pronouns, which are excluded on different grounds, definite descriptions are excluded from the substitution class in (Sub) on the ground that the logical form of an identity sentence flanked by a description (used descriptively) is not given by '$x = y$'. Supporting this analysis are the reasons Russell laid out so meticulously. An identity sentence, true or false, presumes reference, and descriptions need not refer (be satisfied). Further reasons are entailed by the nature of the identity relation itself, which is the relation between an object and an object just in case it is the same object. That an object is the same as itself is not a contingent matter, but that one object uniquely satisfies two sets of properties may well be. We are not propounding the theory of descriptions as *the* analysis of all "the" phrases. There are other and perhaps better ones, in particular, analyses that deal with abstract objects such as numbers. But, whatever the analysis, it should display the predicative role of descriptions as ordinarily used, which is not revealed in the surface grammar of an identity sentence flanked by a definite description.

To specify the substitution class as the class of proper names is not merely to specify a syntactic category. To count as a proper name an expression must refer without being tied to any particular characterization of the object. It requires a naming episode and conditions for the name to enter into the common language as do other words in common speech. Proper names have a logically irreducible use. They permit us to entertain a separation in language of the object under discussion from its properties. How a word, a name, can come to have such a use has been articulated by Keith Donnellan, Saul Kripke, and others.[8] Peter Geach puts it succinctly as follows:

> For the use of a word as a proper name there must in the first instance be someone acquainted with the object named. But language is an institution, . . . and the use of a given name for a given object, . . . like other features of language, can be handed on from one generation to another; . . . Plato knew Socrates, and Aristotle knew Plato, and Theophrastus knew Aristotle, and so on in apostolic succession down to our own times; that is why we can legitimately use 'Socrates' as a name the way we do. It is not

---

8. S. Kripke, "Identity and Necessity," in *Identity and Individuation,* ed. M. Munitz (New York: New York University Press, 1971), pp. 135–164. Also, idem., "Naming and Necessity," in *Semantics of Natural Language,* ed. G. Harman and D. Davidson (Dordrecht: Reidel, 1972). Donnellan, "Reference and Definite Descriptions."

our knowledge of this chain that validates our use, but the existence of such a chain. When a serious doubt arises . . . whether the chain does reach right up to the object named, our right to use the name is questionable, just on that account. But a right may obtain even when it is open to question.[9]

In addition to a specification of the substitution class as proper names, what further condition must be placed on (Sub)? As noted above, the counterexample posed by (1) and (2) is not to be accounted for by a failure to use proper names or by an incomplete analysis of the identity sentence 'Georgione = Barbarelli'. At this juncture Cartwright's analysis is suggestive of additional constraints to place on (Sub) if it is to reflect Leibniz's law, that if *a* and *b* are identical, whatever is true of *a* is true of *b*. Cartwright notes that what is true of Georgione, i.e., being so called because of his size, is not "detachable." Replacing the names by place markers doesn't "capture," so to speak, the predicate. When a sentence contains proper names properly used, properties or relations are ascribed to the objects named, and a representation of such a sentence in logical form should articulate the partition. Displaying form is not like mapping reality. Nor is it very helpful to speak of logical form as "displaying the form of the facts." Logical form is, rather, a way of saying what we want to say: unambiguously, minimizing dependence on nonverbal context for interpretation, and preserving principles of logic and inference. The failure of a sentence to support substitution of the names of identified objects is a measure of the deficiency of that sentence with respect to logical form. The principle of substitutivity may therefore be perceived as both true and regulative. Let us restate it as follows:

(Sub′) For all proper names α and β (indexed to preserve univocality), $\ulcorner \alpha = \beta \urcorner$ expresses a true proposition just in case, for all sentences *P, S,* and *S′*, if *S* is a restatement of *P* in logical form and if *S′* is like *S* save for containing an occurrence of β where *S* contains an occurrence of α, then *S* expresses a true proposition only if *S′* does also.

In proposing (Sub′) we are not also supposing that there is one and only one transformation of a sentence into logical form, for as we noted above, inextricable from all such proposed transformations is

9. P. T. Geach, in *Logic Matters* (Berkeley and Los Angeles: University of California Press, 1972), p. 155. A reprinting of "The Perils of Pauline," *Review of Metaphysics,* XXIII, 2 (1970).

the ontological perspective from which they are recommended. I'm thinking here, for example, of the various solutions that have been offered for the puzzle about 9 and the number of planets. An analysis of

(4)   9 is the number of planets

will depend partially on how one views numbers. Are they sets, properties, or first-order individuals? How one interprets the 'is' of (4) will be guided by such commitments. The familiar solution to the puzzle about substitution failure in modal contexts of '9' and 'the number of planets' takes it that while '9' names an individual the surface-grammar identity sign is flanked by a description in '9 = the number of planets' and (Sub'), therefore, does not apply. Analysis in accordance with the theory of descriptions dispels the substitution failure. But there are alternative analyses. One of them takes it that 9 is a property of sets, and (4) is a deviant way of saying that there are 9 planets. On the latter analysis, although it is necessary that a set with 9 members have more than 7 members, it does not follow that there are necessarily more than 7 planets, since it is a contingency that there are nine of them. The chain of sentences that reflects such an analysis will not generate a substitution failure. But the ultimate ground for choosing between such alternative analyses is not yet resolved. The absence of an account of names that is more adequate to expressions like numerals weighs against the first alternative, but not decisively.[10]

A belief in the principle of substitutivity is grounded in the belief that the pursuit of logical form is not futile. Proper analysis has in fact dispelled apparent failures of substitution in modal contexts, but other contexts, such as epistemological contexts, remain insufficiently analyzed. Nevertheless, as Fitch suggests, it still remains *reasonable* to suppose that the names of identified objects "should be everywhere intersubstitutable where they are being used as names."

10. Kripke and Geach's accounts of how proper names can come to have the use they do raise questions about the nature of those objects about which we say, without further analysis, that an identity relation holds. If the requirement on identity statements is acquaintance with the objects named and that there have been a naming episode presupposes that we are dealing with concrete individuals publicly displayable, then the question arises whether expressions used to designate abstract objects, whatever the syntactic simplicity of those expressions, are being used as proper names. Positing an archetypal white queen (of chess) or a number is not at the same time positing an object that can be properly named in an ordinary way.

8. Nominalism and the Substitutional Quantifier

This paper was a contribution to the *Monist,* LXI, 3 (1978), pp. 351–362. The theme of that issue was nominalism. ■

It has been suggested that a substitutional semantics for quantification theory lends itself to nominalistic aims. I should like in this paper to explore that claim.

Debates about nominalism and realism can be seen in relief against background theories about the relation of a language to the objects the language purports to be about. Confining ourselves to those parts of a language that are vehicles for truth claims—complete sentences, or statements if you like—the supposition of nominalists and realists alike is that there are, in any meaningful statement, links between some or all of the words in the statement, taken singly or in concatenation, and objects that the statements are about. Statements contrive somehow to mention or refer, directly or obliquely, to objects. Plato,[1] in arguing against the Sophists' claim that erroneous beliefs are not about anything, says, "Whenever there is a statement it must be about something," and that, he claimed, holds for false statements as well as true ones.

Statements, true or false, speak of objects. In this, nominalists and realists are generally in agreement. The disagreement on the metaphysical side is about which objects: What exactly is being mentioned, directly or indirectly, by a statement? The disagreement on the linguistic side is about which words are doing the mentioning: Which words in the statement are bearing the burden of reference; all of them, some of them? And, if not all of them, what work is being done by the others? The last question echoes the distinction between categorematic (referring) and syncategorematic (nonreferring) terms.

Historically, the focus of the debate has been the realists' claim that predicate expressions (often broadly conceived to include common nouns and the like as well as verbs) can, with a bit of pushing and shoving, be seen as designating objects: the universals. Among the universals are properties, attributes, relations, or the less abstract, but abstract nevertheless, classes or sets. The realists further claim that those objects are of a categorically different kind, or at least a different species, from the particulars.

Particulars are the concrete individuals that constitute the nominalists' preferred ontology. The relation between particulars and the purported designata of predicates is called—misleadingly, since it seems a grammatical term—"predication." Historically, as we know, there is considerable variation as to what counts as a concrete individual. Nominalists have made varying claims, some of them reduc-

---

1. Plato, *The Sophist*, 262E, in F. M. Cornford, *Plato's Theory of Knowledge* (London: K. Paul, Trench, Trubner & Co., 1935).

tionist, as to whether there are basic particulars, such as events, physical atoms, sense data, time slices of physical objects, out of which other particulars are constructed. But in all its vicissitudes, the primary motivation of nominalism in virtually all its manifestations has not been merely an insistence on the one category of individuatable objects. There is the empirical thrust. The nominalists' individuals are of a kind that can be confronted or that, in the least, make up such confrontable or encounterable individuals. They can, so to speak, put in an appearance. Encounterability by the mind's eye is not generally counted in the spirit of nominalism.

The empirical momentum of nominalism carries it beyond the disclaimer about universals. The nominalist sets himself the task of arguing that abstract "entities" such as numbers, minds, propositions, even where categorically construable as individuals, are in the least otiose or, alternatively, not the abstract objects they are claimed to be. Numbers, for example, may be seen as constructible out of inscriptions; propositions may be seen as an outcome of confusion of material and formal mode, and the like. In summary, nominalism has traditionally moved along two tracks. First, there are not two sorts of things, universals and particulars, that bear an irreducible relation to each other. Second (the empirical thrust), individuals are in one way or another held to be encounterable objects. A philosopher who moves along either track designates himself as nominalistically inclined.

We noted above that nominalist-realist debates will be reflected in accounts of which expressions in a statement bear the burden of reference. A philosopher whose ontology includes universals, as well as propositions, numbers, and other abstract objects, must articulate which components of a statement mention such objects. The philosopher who eschews such putative objects must explain how those components, if they do not refer, are to be understood. How do they work? What do they contribute to the meaning, the content, or, more modestly, the truth conditions of the statement? Displaying the referential apparatus of a statement often takes the form of providing an "analysis"; e.g., replacing the statement by one that is arguably equivalent but that is said to display or articulate better the locus of reference. Russell's theory of descriptions and Quine's conversion of proper names into predicates are two related examples.

If it is names that are the vehicles of reference, then nominalist-realist disagreements will be reflected in what is to be counted as a name. The tendency has been to call any expression a name, however distant from the familiar grammatical category of nouns, provided it is seen as referring. Similarly, genuine name status is denied to terms that are relieved of the burden of reference, however they may have been classified grammatically. Quine,[2] for example, denies that any nouns need be construed as names, proper or otherwise.

Setting aside questions of origin, whether it is someone's convictions about ontology, meager or ample, that lead him to view some words or concatenations of words as referring names, or whether it is the surface grammar that compels the ontology, or the interplay between them, it is surely the case that there are few words or word sequences about which there has been uniform agreement as to whether they refer; whether "logical grammar" reveals them as names. There are those who have claimed that a sentence names a proposition or that it names a possible state of affairs. For Frege, a sentence is contextually ambiguous, naming the true or the false in *oratio recta* and a thought or a proposition in *oratio obliqua*. Quantifiers such as 'nothing' have been interpreted as naming nothingness (*vide* Husserl) or the set of nonexistent "objects." The more traditionally syncategorematic expressions such as 'or' and 'and' have not always been spared reference. 'Not' is seen as naming the falsity property or, more constrainedly, the set of false propositions. 'Or' names alternativeness or the set of pairs of propositions at least one of which has the truth property, and so on. Being itself has been viewed as referent of the verb 'to be'.

There has been more general agreement about the role of proper names, names of individuals, as reference bearers, but here too there are demurrers. It has been claimed that to be a genuine proper name, the name *must* refer, for otherwise it is without meaning, and any sentence of which it is a part would not be sufficiently definite to warrant its having a truth value. Russell, noting that there are meaningful statements with vacuous proper names, argues that it is not ordinary but, rather, "logically" proper names that refer, and those proper names defy incorporation into any common tongue. [A position he later abandoned.] Quine, in one of his philosophical stances, assimilates proper names to predicates altogether. The word that does the referring to Socrates in 'Socrates is a philosopher' does not even occur in the surface structure of that reparsed sentence. Its form is made explicit by the use of quantifiers and variables, as in 'There is something that Socratizes and is a philosopher'. Nevertheless, for the nominalist with an ontology of empirically distinguishable objects, proper

---

2. *Philosophy of Logic* (Englewood Cliffs, N.J.: Prentice-Hall, 1970), pp. 22–30. For Quine, variables bear the burden of reference.

names are seen as a primary vehicle for reference. Their objects are of the kind that, at least theoretically, can be properly named by an act of ostension.

Let us suppose that the realists we are juxtaposing against the nominalists are not of the most lavish kind. Both camps agree that there are concrete individuals that are not predicated of anything. The realist also admits referents of predicates: universals (or, more modestly, sets) that are of a different category in that the relationship between an individual and a universal is not eliminable or reducible to a relation among individuals. The realist generally holds the relation of predication to be irreflexive and asymmetric. If the realist claims, as he usually does, that properties and relations may themselves have properties and enter into relationships, then, if the relation of predication is taken as univocal, it is generally held to be intransitive: a hierarchical ordering with individuals at the base—the zero-level objects.

A major task of nominalists is then to explain how predicates work. In recent years, attempts to give a formal account of nominalism have gone in two directions. The first is to reconstruct the predication relation as one that holds between individuals. The second is to deny altogether the referential function of predicates; predication ceases to be understood as a relation between objects.

The first kind of reconstruction may take different forms.[3] One is to conceive of a predicate as having divided reference. Let '*S*' name some individual and let '*P*' be a monadic predicate, then '*S* is *P*' is interpreted as saying that *S* is identical with one of the referents of *P*. '*P*' does not name a set here. The relation of predication is that of identity with one of the referents of the predicate. There is no new object for the predicate to name. Another reconstruction that reduces predication to a relation among zero-level individuals is to view it as inclusion between parts and wholes, all of which are individuals and where the whole is generated by its ultimate parts. The part-whole analysis can take two forms. In '*S* is *P*', *S* may count as the part with *P* as the whole; this is analogous to some interpretations of mass nouns where, for example, 'water' is viewed as the name of an individual made up of dispersed individual parts. An alternative is to count *P* as the part and *S* as the whole—the case where, for example, individuals are seen as bundles of qualities and it is the qualities that are the ultimate individual parts.

---

3. For an excellent account of such reductionist nominalistic programs, see R. A. Eberle, *Nominalistic Systems* (Dordrecht: Reidel, 1970).

A further reductive alternative is to take the relation of predication as similarity and '*P*' as naming some selected individual that serves as a standard. Predication is then likeness with respect to some feature of *S* and some feature of *P*.

In the foregoing proposals, the predicate '*P*' is claimed to refer, whether singly or multiply. The relation of predication holds between individuals of the same type level, whether it be similarity, identity, or part-whole. Contrasted with such reductive theories is the other direction a nominalist reconstruction can take. Predication is not seen as a relation between objects. It is viewed as a grammatical construction that generates open sentences from predicates and variables when arranged in proper order, as in '*x* flies' or '*x* fears *y*'. Those constructions can be understood by an English speaker without there being something to be named by 'fears' or 'flies'. The open sentences generated by predication are satisfied or not satisfied by individuals or sequences of individuals. They are, speaking elliptically, true of the objects that satisfy them.

It is not my purpose to evaluate the success of the various nominalistic enterprises. Indeed, I do not see the proposals as necessarily competitive, since there is no reason to suppose that predication demands a uniform analysis. In the present paper, I will be concerned primarily with the last-mentioned alternative or weakly realist modifications of it: I will be considering either the no-name theory of predication, which views predicates as akin to the syncategorematics, or weakly realist extensions, which view predicates as naming sets of individuals and predication as a relation of membership.

Names of individuals have traditionally been assigned an important referential role, in contrast to the nonreferential role that nominalism has traditionally tried to assign to expressions that do not name particular individuals. Proper names were viewed as linguistic ''signs'' for concrete, particular, encounterable objects. Hobbes saw proper names as linguistic conventions devised to function like natural, nonlinguistic signs of actual objects. With the development of formal logic and its standard semantics, proper names have taken on a more subsidiary role, even for those with nominalist inclinations.

It is important for the subsequent discussion of the relation between the substitutional quantifier and nominalism to consider the grounds for the shift in the role assigned to proper names. There is the obvious fact that, of the nominalists' nameable objects, very few are actually named. Meaningful sentences, true or false, which are about such objects, may contain no names at all; an example is 'Something is green'. However, implicit in even medieval nominalism[4] is the suggestion that the absence of actual names is accidental. The medieval nominalist's objects are in principle nameable, and (the meanings of) sentences like 'Something is green' are taken as abbreviating a finite disjunction of singular sentences where expressions in subject position are proper names. (Also, implicit here is that names, given that they are conventional signs, are theoretically always available.) Similarly, the universal quantifier is interpreted as abbreviating a conjunction of singular sentences. In modern quantification theory, it was not the absence of actual names but, rather, the possibility of an infinity of objects that was the catalyst for introducing quantifiers as primitive logical operators. An adequate language for referring to infinitely many objects would seem to require variables and quantifiers in addition to names.

But others have pressed even further. Do we need proper names of individuals at all? Can variables and quantifiers be made to bear the entire burden of reference? The suggestion is already manifest in Russell's theory of descriptions. Russell noted that singular descriptions function grammatically (in surface grammar) like ordinary proper names. But there are vacuous descriptions, such as 'the present king of France', and since, for Russell, *genuine* names must refer, singular descriptions are not genuine names. Yet statements in which vacuous descriptions occur are not incomplete; they have a truth value. So there must be a way of unpacking statements with singular descriptions that preserves their 'content' and in which singular descriptions are seen not to be genuine names. The theory of descriptions provides the mode of translation. 'The present king of France is bald' becomes roughly 'There is something and just one, such that it is a present king of France and bald'. Unpacked, the statement contains quantifiers and other logical operators, variables, and predicates. Russell, without abandoning proper names as a vehicle for reference altogether (he retained the elusive "logically" proper names) does go on to say that what prompts the analysis for singular descriptions seems also to prompt it for ordinary proper names. If it makes sense, as it seems to, to ask whether Homer exists, then there are vacuous ordinary proper names. Ordinary proper names must therefore also fail to be genuine proper names, and Russell sees them as classifiable along with the singular descriptions. He called them "truncated descriptions." Whether Russell is claiming that we can select out some particular

---

4. See Ernest A. Moody, "Medieval Logic," in *Encyclopedia of Philosophy* (New York: Macmillan, 1967), ed. P. Edwards, vol. 4, pp. 530–531.

descriptive characterization of which the ordinary proper name is an abbreviation is not of concern to us here. Once the similarity is noted, the analysis can be pressed. On the theory of descriptions, descriptive phrases work like predicates, and the most straightforward recognition of that logical role for ordinary proper names is to convert names directly into predicates. Hence Quine's unabashed proposal that we transform ordinary proper names into verbs in our formal or "regimented" language. 'Pegasus is a winged horse' goes into 'There is something and only one such that it Pegasizes, is winged, and is a horse'. As in the case of vacuous descriptions, the latter turns out to be false. Pursuing this analysis, Quine[5] calls proper names, even where they name individuals that are in a domain of genuine objects, "a frill." They are, he says, "a mere convenience and strictly redundant."

It should be noted that, whereas the no-name theory of *predicates* is consistent with nominalist aims, the redundancy theory of proper names *does,* in my view, represent a departure from traditional nominalism. It is true that the redundancy theory of proper names claims not that proper names do not refer but only that the work can be better done by quantifiers and variables. Still, the redundancy theory does claim *priority* for quantifiers and variables as the machinery for reference. In this there is a considerable departure from traditional nominalism, where the role of names as conventional signs for objects has been central to the theory of reference. I want to claim that one of the connections that hold between substitutional quantification and nominalism is that, in substitutional theory, names are not frills. They are not redundant and eliminable. The juncture at which substitutional theory and the objectual theory of quantifiers overlap is where the substitution class consists of genuine proper names: where the names link up with objects.

Consider a standard first-order language without identity. It has a denumerable stock of variables, individual constants, and predicate letters for *n*-adic predicates. It contains sentential connectives and quantifiers. Formulas are defined inductively. Closed formulas are sentences. Individual constants correspond to proper *names*. A standard interpretation of such a language is given roughly as follows. A domain of individuals is specified over which variables and quantifiers are said to 'range.' Individual constants do not range but name specific objects. On the no-name theory of predication, predicates name nothing. On the more standard interpretations, sets of objects are assigned

5. *Philosophy of Logic*, p. 25.

to predicates: just those sets of *n*-tuples that satisfy the open atomic formula containing the predicate. Truth is defined for sentences of the language via the semantical notion of satisfaction of a formula by a sequence of objects. The procedure is familiar, and we need not spell it out.[6] Given the inductive definition of satisfaction of a formula relative to a sequence of objects, the definition of truth is derivative. It is a property of sentences. A sentence is true just in case it is satisfied by some sequence. What should be noted is that variables do not function as mere place markers. They take objects as values relative to a sequence of objects. Since satisfaction of quantified formulas is defined relative to objects in the domain, they can be read with existential import: "There is something such that. . . ." "Everything is such that. . . ."

On a substitutional semantics of the same first-order language, a domain of objects is not specified. Variables do not range over objects. They are place markers for substituends. Satisfaction relative to objects is not defined. Atomic sentences are assigned truth values. Truth for sentences built up out of the sentential connectives are defined in the usual way. The quantifier clauses in the truth definition say that

$(x)Ax$ is true just in case $A(t)$ is true for all names t.

$(\exists x)Ax$ is true just in case $A(t)$ is true for at least one name t.

In the above summary characterization of a substitutional semantics the substitution class is taken as the class of proper names: just those syntactical expressions that on the referential [objectual] interpretation might be used to refer univocally to objects. However, given the detachment of the substitutional quantifier from reference, and given the truth definition in terms of replacement of variables by expressions, substitution classes (and associated quantifiers) can be expanded. Predicates and sentences can, *inter alia,* also compose substitution classes. If sentences are to be included, inductive requirements constrain permissible substitutions to those that reduce complexity.[7] The same would hold had individual descriptions been included in the substitution class of names.

Before proceeding to the question of the relevance of substitutional semantics to nominalistic aims, a word about further comparisons between the alternative interpretations. Saul Kripke, in an important pa-

6. See, for example, Quine, *Philosophy of Logic,* pp. 35–46. Also, Saul Kripke, "Is There a Problem about Substitutional Quantification?" in *Truth and Meaning,* ed. G. Evans and J. McDowell (Oxford: Clarendon Press, 1976), p. 328.

7. See my "Quantification and Ontology," this volume.

per,[8] has given a detailed account of the substitutional quantifier. He sees the substitutional quantifier as appended to a base language, all of whose sentences are taken as atomic relative to that extension. His generalization allows, as a base language, a first-order language with referential quantifier. The substitutional first-order language presented in the present paper may be viewed as a minimal extension allowed by Kripke's more general characterization, the base language being one that contains no quantifiers.[9] What Kripke proposes is that the substitutional quantifier is not a replacement for, or in competition with, the standard interpretation. If notational differences that distinguish both kinds of quantifiers and variables are preserved, the two kinds can occur concurrently and coherently in the extended language. I have a different view of the matter. I see the substitutional account as the more general. Of the many substitution classes for which the quantifier may be defined, one contains expressions that are vehicles for direct reference: the grammatically proper names. That names may have such a special role with respect to encounterable objects seemed to me apparent and has been argued elsewhere.[10] *How* they come to such a role has been illuminated by the causal theory of names elaborated by Keith Donnellan, Kripke,[11] and others. Proper names, unlike definite descriptions, can be used by speakers to refer to objects without mediation of ''concepts'' or descriptive clusters. They may be used to capture and institutionalize an act of ostension.

Now, it is perfectly clear that we have no such complete stock of actual names, nor must we have. A substitutional first-order language is after all formal, regimented, and theoretical. It supposes initially a denumerably infinite set of names, and we have the expressive wherewithal for generating them. Now, suppose each of our denumerably infinite stock of names does refer to an object. Let those objects make up our reference class: a domain. Under those conditions, we can introduce a substitutional analogue of satisfaction of a formula relative to that domain. If all the contexts of the interpreted language are transparent, then the substitutional analogue of satisfaction converges with the referential definition of satisfaction. Under those con-

8. ''Is There a Problem about Substitutional Quantification?''

9. For a more complete account of a minimal substitutional semantics, see J. M. Dunn and N. D. Belnap, ''The Substitution Interpretation of the Quantifiers,'' *Noûs,* II (1968): 177–185.

10. See my ''Modalities and Intensional Languages,'' this volume.

11. See Saul Kripke, ''Identity and Necessity,'' in *Identity and Individuation,* ed. M. Munitz (1971), pp. 135–164, and ''Naming and Necessity,'' in *Semantics of Natural Language,* eds. D. Davidson and G. Harman (1972). Also K. Donnellan, ''Proper Names and Identifying Descriptions,'' in *Semantics of Natural Languages,* pp. 356–379.

ditions the quantifiers can be read with existential import. I see the referential quantifier as a limiting case. Substitutional quantification, together with a substitution class of names that define a reference class of objects, yields a referential quantifier. If our substitutional language allows wider substitution classes beyond the set of referring names, then of course it is important to distinguish with an alternative notation those cases of quantification where substituends are referring names, for it is *those* cases that can be read back into English as 'There is something such that' and 'Everything is such that'. They have existential import.[12]

I should like now to discuss the ways in which a substitutional interpretation of the quantifier lends itself to nominalistic aims. That it does so lend itself is apparent, but that aspect has perhaps been misconstrued. Quine,[13] who originally regarded the substitutional interpretation of the quantifier as incoherent, with use mention confusions lurking behind it, has over the years, given his nominalistic dispositions, come to notice its appeal. In his speculative theory of language learning in the recent *Roots of Reference* he sees it as reflecting early language acquisition, and he goes on to attempt a substitution interpretation of sets the ''existence'' of which he describes as a major concession to realism. He believes it fails for sets and is disappointed. Others have viewed substitutional theory with alarm because, given that ''To be is to be the value of a bound variable'' and ''Existence is what existential quantification expresses,'' it would seem that, upon translation into a substitutional language, ordinary English sentences like 'There are philosophers' would be divested of their existential import. But that of course supposes that the quantifiers are always the primary locus of reference. There are alternative analyses for locating references in an interpreted language. Names and their relation to *nameable* objects constitute one such alternative. The burden of reference is shifted back univocally to the name relation. As I argued in an earlier paper,[14] where the subject matter is well defined, ''where we are *already* ontologically committed . . . then, all right: to be is to be the value of a variable. If we already believe

12. If our substitution class of names is extended to include nonreferring (syntactic) names as well, then it is for the denumerable subset of referring names that the quantifiers are read with existential import.
13. W. V. Quine, ''Reply to Professor Marcus,'' in *Ways of Paradox* (New York: Random House, 1966). The shift in Quine's view of substitutional quantification can be traced through ''Ontological Relativity,'' *Journal of Philosophy*, LXV (1965): 185–212; *Ontological Relativity and Other Essays* (New York: Columbia University Press, 1969); and *Roots of Reference* (La Salle, Ill.: Open Court, 1973).
14. ''Quantification and Ontology,'' this volume.

. . . in the existence of physical objects . . . then if in our interpretation [in a first-order language] physical objects . . . turn up as
objects over which variables range, this squares with the [ontological]
status they have *already* been granted.'' But if the standard semantics
of first-order logic is taken as paradigmatic, it also *inflates* ontology
or, alternatively, places rigid constraints on the formal languages
available for paraphrase. Consider first opaque contexts. 'Hesperus is
Phosphorus' and 'John believes that Hesperus is Hesperus' can be true
while 'John believes that Hesperus is Phosphorus' might be false.
Given that substitutivity of identity holds in referential first-order languages with identity, such direct translation would require, as in
Frege, that, in opaque contexts, names refer to something other than
their ordinary referents, such as the *senses* of the names. On the substitutional view, there is nothing incoherent about such failures of substitutivity even where the names 'Hesperus' and 'Phosphorus' refer
to the same physical object. We need not inflate ontology with elusive
abstract objects such as senses.

The nominalist finds that standard semantics shackles him to first-
order languages if, like a good nominalist, he is to make do without
abstract higher-order objects. Substitutional semantics permits quantifiers with predicates as substituends without a *prima facie* presumption of reference to universals. Of course in those instances the
quantifier paraphrases cannot be 'There exist . . .' or 'Everything is
such that . . .' and kindred locutions. There are, even in ordinary
use, quantifier phrases that seem to be ontologically more neutral, as
in "It is sometimes the case that species and kinds are, in the course
of evolution, extinguished". It does not seem to me that the presence
there of a quantifier *forces* an ontology of kinds or species. If the case
is to be made for reference of kind terms, it would have to be made,
as for proper names, independently. Translation into a substitutional
language does not force the ontology. Such usage remains, literally
and until the case for reference can be made, *a façon de parler*. That
is the way the nominalist would like to keep it.[15]

Particularly illuminating is the case of higher-order quantification
where variables take sentences as substituends. On a referential theory,

---

15. Kripke has suggested in "Naming and Necessity" that species names and kind
names are proper names of abstract objects, i.e., essences. This is a case where a nominalistic reduction with predication taken as similarity with respect to structure seems to
me more appropriate with some initial sample taken as standard. Not because such a reduction is a desideratum in all cases of predication but because it fits the common-sense
picture for species and kind terms. Structural features of things count as essential properties.
Such properties can be characterized without at the same time requiring that there be essences. See my "Essential Attribution," this volume.

such quantification commits us to designata for sentences. Are they propositions, thoughts, states of affairs, facts? Problems associated with each of those alternatives have afflicted philosophy. What are the identity conditions for such objects, how are they individuated? If the objects are thoughts (not in Frege's sense), aren't we back to psychologism? If the objects are facts or states of affairs, there is the nagging business of negative facts, negative states of affairs, and so on. Not so for the substitutional semantics. On such a view, to have quantified sentences with sentences as substituends is very natural. '$(p)(p \lor \sim p)$' need not be paraphrased as 'Any proposition bears the excluded middle relation to its negation'.

Indeed, on the substitutional view we can in a natural way quantify in and out of quotation contexts. Expressions, unlike nonlinguistic objects, lend themselves to being named by a simple function like a quotation function that at the same time displays the object named. On the substitutional view we can define quantification where substituends are sentences without generating a liar paradox. As I showed elsewhere,[16] general inductive constraints that are required of any definition, when applied to the definition of the quantifiers, thwart the antinomy. All this without introducing propositions or other elusive referents for sentences—a great comfort to the nominalist.

Also consistent with nominalist aims is the fact that the substitutional view articulates a distinction between truth and reference, with the burden of reference to be borne by just those expressions that do in fact refer—the names of individuals. If nominalism is linked to empiricism, as it usually is, individuals are in principle confrontable, encounterable, dubbable by an act of ostension, or, in the least, are made up of or make up such encounterable objects. Since we encounter so few objects and name even fewer, it seems there is a gap to be filled. How any name can have this ostensive role has been, as we noted above, illuminated by the causal theory of names. What the nominalist further requires is that individuals be such that they are, theoretically, properly nameable. For domains so defined, the substitutional quantifier can be read referentially.

It is important to notice again that the relation of identity, holding as it does between individuals, cannot sensibly be introduced into the substitutional language except where the expressions flanking the identity sign are genuine proper names of individuals. Nor will substitutivity always hold for identity. Conversely, substitutivity *salve veritate* cannot define identity, since two expressions may be everywhere

16. "Quantification and Ontology," this volume.

intersubstitutable and not refer at all. Identity and substitutivity converge where there is reference and transparency of context up to and including modal contexts.[17]

Critics of substitutional semantics have pointed out that, if there are nondenumerably many objects, there might, on the substitutional view, be true universal sentences that are falsified by an unnamed object, and there must always be some such, for names are denumerable. But the fact that every *referential* first-order language that has a nondenumerable model must have a denumerable model gives little advantage to the referential view. There are, however, other reasons that leave the nominalist unperturbed. The nominalist has always been diffident about nondenumerable collections, and he looks to alternative accounts of, for example, the real numbers. He is even suspicious of denumerable infinities, but sees them as a more natural extension in much the same way that the substitutional quantifier is a natural extension of conjunction and disjunction for a denumerably infinite substitution class.

What the nominalist might look for in the way of an ideal formal language is[18] type-theoretic with a denumerable stock of proper names at level zero. All of those names or some denumerably infinite subset thereof may be taken as referring to nominalistically acceptable objects. At lowest level, where substituends are referring names, the quantifiers may be read existentially. Beyond lowest level, the variables and quantifiers are read substitutionally. There need be no new objects of reference for higher-order expressions like predicates, sentences, sentential connectives.

Now, it is not at all clear that such a program is wholly feasible, but it is surely a nominalistic program.

---

17. See Kripke, ''Is There a Problem about Substitutional Quantification?'' §§3, 6.
18. See ibid., p. 368.

9. *Moral Dilemmas and Consistency*

This paper was published in the *Journal of Philosophy*, LXXVII, 3 (March 1980): 121—136. It has a protracted history. One of its central claims is that, where moral conflict in a particular case derives from one or more general principles, that fact is evidence neither for the inconsistency of the principles nor for the inconsistency of the particular judgments that flow from those principles, although the actions mandated may be incompatible. I first presented some of my views in a symposium with John Lemmon sponsored by the American Philosophical Association in May 1964. The arguments seemed obvious to me, so I assumed they were obvious to all and filed the paper away.

But the notion persisted that moral dilemmas are evidence for inconsistency. In the late 1970s Robert Fogelin, Walter Sinnott-Armstrong, then a Yale graduate student, and I were discussing moral realism, a subject that Sinnott-Armstrong was considering for a dissertation. I pointed out that the inconsistency account of moral conflict could not be sustained and, therefore, does not count as evidence for nonrealism. They urged me to publish my views. Sinnott-Armstrong's dissertation evolved into a comprehensive study of moral conflict (*Moral Dilemmas* [Cambridge: Blackwell, 1988]) that goes far beyond the issues here discussed.

My claims converge with some of Bernard Williams's views at several junctures, such as the contingent origin of moral conflict, the psychological residue in the case of nonselected alternatives, and the failings of some extant deontic logics to capture the "moral facts." But we differ on central issues. Williams says, "The nonrealist approach may well allow for the possibility that one can be forced to two *inconsistent* [italics mine] moral judgments about the same situation, each of them backed by the best possible reasons . . ."—an alternative that he claims is not open to the moral realist (*Problems of the Self* [New York: Cambridge University Press, 1973], p. 205).

As I saw it, the usual cases of dilemma are not cases of inconsistent moral judgments and hence not evidence against moral realism.

The paper was completed during my 1979 stay at the Center for Advanced Study in the Behavioral Sciences, where I profited from discussions with resident fellows and Stanford-area philosophers. It was also presented at several departmental colloquia and as the Gail Stine Memorial Lecture at Wayne State University in 1980.

A footnote to footnotes 9 and 11 below: As I indicated, in the case of rejected alternatives, particularly in hard cases, "regret" seems too weak a characterization of the appropriate feeling, and "remorse," with its dimension of guilt, though more appropriate, may mislead. I learned from David Sachs and John Rawls that, in a 1956 seminar at Cornell, Rawls had proposed "regorse" as a term for an intermediate moral sentiment.

I cannot refrain from reporting a student comment on one of the occasions when I read this paper: He said, "I didn't come all the way out to California for you to lay a guilt trip on me." ∎

I want to argue that the existence of moral dilemmas, even where the dilemmas arise from a categorical principle or principles, need not and usually does not signify that there is some inconsistency (in a sense to be explained) in the set of principles, duties, and other moral directives under which we define our obligations either individually or socially. I want also to argue that, on the given interpretation, consistency of moral principles or rules does not entail that moral dilemmas are resolvable in the sense that acting with good reasons in accordance with one horn of the dilemma erases the original obligation with respect to the other. The force of this latter claim is not simply to indicate an intractable fact about the human condition and the inevitability of guilt. The point to be made is that, although dilemmas are not settled without residue, the recognition of their reality has a dynamic force. It motivates us to arrange our lives and institutions with a view to avoiding such conflicts. It is the underpinning for a second-order regulative principle: that, as rational agents with some control of our lives and institutions, we ought to conduct our lives and arrange our institutions so as to minimize predicaments of moral conflict.

I

Moral dilemmas have usually been presented as predicaments for individuals. Plato, for example, describes a case in which the return of a cache of arms has been promised to a man who, intent on mayhem, comes to claim them. Principles of promise-keeping and benevolence generate conflict. One does not lack for examples. It is safe to say that most individuals for whom moral principles figure in practical reasoning have confronted dilemmas, even though these more commonplace dilemmas may lack the poignancy and tragic proportions of those featured in biblical, mythological, and dramatic literature. In the one-person case there are principles in accordance with which one ought to do x and one ought to do y, where doing y requires that one refrain from doing x. For the present rough-grained discussion, the one-person case may be seen as an instance of the n-person case under the assumption of shared principles. Antigone's sororal (and religious) obligations conflict with Creon's obligations to keep his word and preserve the peace. Antigone is obliged to arrange for the burial of Polyneices; Creon is obliged to prevent it. Under generality of principles they are each obliged to respect the obligations of the other.

It has been suggested that moral dilemmas, on their face, seem to reflect some kind of inconsistency in the principles from which they derive. It has also been supposed that such conflicts are products of a plurality of principles and that a single-principled moral system does not generate dilemmas.

In the introduction to the *Metaphysics of Morals* Kant[1] says, "Because however duty and obligation are in general concepts that express the objective practical necessity of certain actions . . . it follows . . . that a conflict of duties and obligations is inconceivable *(obligationes non colliduntur)."* More recently John Lemmon,[2] citing a familiar instance of dilemma, says, "It may be argued that our being faced with this moral situation merely reflects an implicit inconsistency in our existing moral code; we are forced, if we are to remain both moral and logical, by the situation to restore consistency to our code by adding exception clauses to our present principles or by giving priority to one principle over another, or by some such device. The situation is as it is in mathematics: there, if an inconsistency is revealed by derivation, we are compelled to modify our axioms; here, if an inconsistency is revealed in application, we are forced to revise our principles." Donald Davidson,[3] also citing examples of conflict, says, "But then unless we take the line that moral principles *cannot* conflict in application to a case, we must give up the concept of the nature of practical reason we have so far been assuming. For how can premises, all of which are true (or acceptable), entail a contradiction? It is astonishing that in contemporary moral philosophy this problem has received little attention and no satisfactory treatment."

The notion of inconsistency that views dilemmas as evidence for inconsistency seems to be something like the following. We have to begin with a set of one or more moral principles that we will call a *moral code.* To count as a principle in such a code, a precept must be of a certain generality; that is, it cannot be tied to specific individuals at particular times or places, except that on any occasion of use it takes the time of that occasion as a zero coordinate. The present rough-grained discussion does not require that a point be made of the distinction between categorical moral principles and conditional moral principles, which impose obligations upon persons in virtue of some

1. Immanuel Kant, *The Metaphysical Elements of Justice,* Part I of the *Metaphysics of Morals,* trans. John Ladd (Indianapolis: Bobbs-Merrill, 1965), p. 24.

2. "Deontic Logic and the Logic of Imperatives," *Logique et analyse,* VIII, 29 (April 1965): 39–61, a revision of a paper presented at a symposium of the Western Division meeting of the American Philosophical Association in May 1964. My unpublished comments on that occasion contain some of the ideas here advanced.

3. "How Is Weakness of the Will Possible?," in *Moral Concepts,* ed. Joel Feinberg (New York: Oxford University Press, 1970), p. 105.

role, such as that of being a parent or a promise-maker or contractee. For our purposes we may think of categorical principles as imposing obligations in virtue of one's being a person and a member of a moral community.

In the conduct of our lives, actual circumstances may arise in which a code mandates a course of action. Sometimes, as in dilemmas, incompatible actions x and y are mandated; that is, the doing of x precludes the doing of y; y may in fact be the action of refraining from doing x. The underlying view that takes dilemmas as evidence of inconsistency is that a code is consistent if it applies without conflict to all actual—or, more strongly—to all possible cases. Those who see a code as the foundation of moral reasoning and adopt such a view of consistency argue that the puzzle of dilemmas can be resolved by elaboration of the code: by hedging principles with exception clauses, or establishing a rank ordering of principles, or both, or a procedure of assigning weights, or some combination of these. We need not go into the question of whether exception clauses can be assimilated to priority rankings, or priority rankings, to weight assignments. In any case, there is some credibility in such solutions, since they fit some of the moral facts. In the question of whether to return the cache of arms, it is clear (except perhaps to an unregenerate Kantian) that the principle requiring that the promise be kept is overridden by the principle requiring that we protect human lives. Dilemmas, it is concluded, are merely apparent and not real, for, with a complete set of rules and priorities or a complete set of riders laying out circumstances in which a principle does not apply, in each case one of the obligations will be vitiated. What is incredible in such solutions is the supposition that we could arrive at a complete set of rules, priorities, or qualifications that would, in every possible case, unequivocally mandate a single course of action; that where, on any occasion, doing x conflicts with doing y, the rules with qualifications or priorities will yield better clear reasons for doing one than for doing the other.

The foregoing approach to the problem of moral conflict—ethical formalism—attempts to dispel the reality of dilemmas by expanding or elaborating on the code. An alternative solution, that of moral intuitionism, denies that it is possible to arrive at an elaboration of a set of principles that will apply to all particular circumstances. W. D. Ross,[4] for example, recognizes that estimates of the stringency of different *prima facie* principles can sometimes be made, but argues that no general universally applicable rules for such rankings can be laid

---

4. *The Right and the Good* (New York: Oxford University Press, 1930), p. 41.

down. However, the moral intuitionists *also* dispute the reality of moral dilemmas. Their claim is that moral codes are only guides; they are not the only and ultimate ground of decision making. *Prima facie* principles play an important role in our deliberations, but not as a set of principles that can tell us how we ought to act in all particular circumstances. That ultimate determination is a matter of intuition, albeit rational intuition. Moral dilemmas are *prima facie,* not real, conflicts. In apparent dilemmas there *is* always a morally correct choice among the conflicting options; it is only that, and here Ross quotes Aristotle, "the decision rests with perception." For Ross, those who are puzzled by moral dilemmas have failed to see that the problem is epistemological and not ontological, or real. Faced with a dilemma generated by *prima facie* principles, *uncertainty* is increased as to whether, in choosing x over y, we have in fact done the right thing. As Ross puts it, "Our judgments about our actual duty in concrete situations, have none of the certainty that attaches to our recognition of general principles of duty. . . . Where a possible act is seen to have two characteristics in virtue of one of which it is *prima facie* right and in virtue of the other *prima facie* wrong we are well aware that we are not certain whether we ought or ought not to do it. Whether we do it or not we are taking a moral risk." For Ross, as well as the formalist, it is only that we may be uncertain of the right way. To say that dilemma is evidence of inconsistency is to confuse inconsistency with uncertainty. There *is* only one right way to go, and hence no problem of inconsistency.

There are, as we see, points of agreement between the formalist and the intuitionist as here described. Both claim that the appearance of dilemma and inconsistency flows from *prima facie* principles and that dilemmas can be resolved by supplementation. They differ on the nature of the supplementation.[5] They further agree that it is the multiplicity of principles that generates the *prima facie* conflicts; if there were one rule or principle or maxim, there would be no conflicts.

Quite apart from the unreasonableness of the belief that we can arrive ultimately at a single moral principle, such proposed single principles have played a major role in moral philosophy. Kant's categorical imperative and various versions of the principle of utility being primary examples. Setting aside the casuistic logical claim that a single principle can always be derived by conjunction from a multiplicity,

5. For the formalist, priority rankings (like Rawls's lexical ordering), or weights permitting some computation, or qualifications of principles to take care of all problematic cases, are supposed possible. For the intuitionist, it is intuitive "seeing" in each case that supplements *prima facie* principles.

it can be seen that the single-principle solution is mistaken. There is always the analogue of Buridan's ass. Under the single principle of promise-keeping, I might make two promises in all good faith and reason that they will not conflict, but then they do, as a result of circumstances that were unpredictable and beyond my control. All other considerations may balance out. The lives of identical twins are in jeopardy, and, through force of circumstances, I am in a position to save only one. Make the situation as symmetrical as you please. A single-principled framework is not necessarily unlike a multiprincipled code with qualifications or priority rules, in that it would appear that, however strong our wills and complete our knowledge, we might be faced with a moral choice in which there are no moral grounds for favoring doing x over y (or vice versa).

Kant imagined that he had provided a single-principled framework from which all maxims flowed. But Kantian ethics is notably deficient in coping with dilemmas. Kant seems to claim that they don't really arise, and we are provided with no moral grounds for their resolution.

It is true that unregenerate-act utilitarianism is a plausible candidate for a dilemma-free principle or conjunction of principles, but not because it can be framed as a single principle. It is rather that attribution of rightness or wrongness to certain *kinds* of acts per se is ruled out whether they be acts of promise-keeping or promise-breaking, acts of trust or betrayal, of respect or contempt. One might, following Moore, call such attributes ''non-natural kinds,'' and they enter into all examples of moral dilemmas. The attribute of having maximal utility as usually understood is not such an attribute, for to the unregenerate utilitarian it is not features of an act per se that make it right. The only thing to be counted is certain consequences, and, for any given action, one can imagine possible circumstances, possible worlds if you like, in each of which the action will be assigned different values—depending on different outcomes in those worlds. In the unlikely cases where in fact two conflicting courses of action have the same utility, it is open to the act utilitarian to extend computation by adopting a procedure for deciding, such as tossing a coin.

In suggesting that in all examples of dilemma, we are dealing with attributions of rightness per se independent of consequences is not to say that unambiguous principles of utility do not enter into moral dilemmas. It is only that such conflicts will emerge in conjunction with nonutilitarian principles. Indeed, such conflicts are perhaps the most frequently debated examples, but not, as we have seen, the only ones. I would like to claim that it is a better fit with the moral facts that all dilemmas are real, even where the reasons for doing x outweigh, and in whatever degree, the reasons for doing *y*. That is, wherever circumstances are such that an obligation to do *x* and an obligation to do *y* cannot as a matter of circumstance be fulfilled, the obligations to do each are not erased, even though they are unfulfillable. Mitigating circumstances may provide an explanation, an excuse, or a defense, but I want to claim that this is not the same as denying one of the obligations altogether.

We have seen that one of the motives for denying the reality of moral dilemmas is to preserve, on some notion of consistency, the consistency of our moral reasoning. But other not unrelated reasons have been advanced for denying their reality that have to do with the notion of guilt. If an agent can and ought to do *x*, then he is guilty if he fails to do it. But if, however strong his character and however good his will and intentions, meeting other equally weighted or overriding obligations precludes his doing *x*, then we cannot assign guilt, and, if we cannot, then it is incoherent to suppose that there is an obligation. Attendant guilt feelings of the agent are seen as mistaken or misplaced.

That argument has been rejected by Bas van Fraassen[6] on the ground that normative claims about when we ought to assign guilt are not part of the analysis of the concept of guilt, for if they were, such doctrines as that of "original sin" would be rendered incoherent. The Old Testament assigns guilt to three or four generations of descendants of those who worship false gods. Or consider the burden of guilt borne by all the descendants of the house of Atreus, or, more recently, the readiness of many Germans to assume a burden of guilt for the past actions of others. There are analogous converse cases, as in the as-

---

6. "Values and the Heart's Command," *Journal of Philosophy*, LXX, 1 (January 11, 1973): 5–19. Van Fraassen makes the point that such a claim would make the doctrine of "original sin" incoherent. As I see it, there are at least three interesting doctrines, one of them very likely true, that could qualify as doctrines of original sin.

One of them, which I call "inherited guilt," is the doctrine that some of the wrongful actions of some persons are such that other persons, usually those with some special connection to the original sinners, are also judged to be sinners; their feelings of guilt are appropriate, their punishment "deserved," and so on. Such is the case described in Exodus and Deuteronomy here mentioned.

A second notion of original sin is to be found in an account of the Fall. Here it is suggested that, however happy our living arrangements, however maximal the welfare state, we will each of us succumb to some temptation. There is universality of sin because of universality of weakness of will, but specific sins are neither inherited by nor bequeathed to others.

A third candidate supposes the reality and inevitability, for each of us, of moral dilemma. Here we do not inherit the sins of others, nor need we be weak of will. The circumstances of the world conspire against us. However perfect our will, the contingencies are such that situations arise where, if we are to follow one right course of action, we will be unable to follow another.

sumption of guilt by some parents for actions of even adult children. Having presented the argument, I am not wholly persuaded that a strong case can be made for the coherence of such doctrines. However, the situation faced by agents in moral dilemmas is not parallel. Where moral conflict occurs, there is a genuine sense in which both what is done and what fails to be done are, before the actual choice among irreconcilable alternatives, within the agent's range of options. But, as the saying goes—and it is not incoherent—you are damned if you do and you are damned if you don't.

I will return to the claim that moral dilemmas are "real," but first let me propose a definition of consistency for a moral code that is compatible with that claim.

II

Consistency, as defined for a set of meaningful sentences or propositions, is a property that such a set has if it is possible for all of the members of the set to be true, in the sense that contradiction would not be logical consequence of supposing that each member of the set is true. On that definition "grass is white" and "snow is green" compose a consistent set although false to the facts. There is a possible set of circumstances in which those sentences are true, i.e., where snow is green and grass is white. Analogously, we can define a set of rules as consistent if there is some possible world in which they are all obeyable in all circumstances in *that* world. (Note that I have said "obeyable" rather than "obeyed," for I want to allow for the partition of cases where a rule-governed action fails to be done between those cases where the failure is a personal failure of the agent— an imperfect will in Kant's terms—and those cases where "external" circumstances prevent the agent from meeting conflicting obligations. To define consistency relative to a kingdom of ends, a deontically perfect world in which all actions that ought to be done are done, would be too strong, for that would require both perfection of will *and* the absence of circumstances that generate moral conflict.) In a world where all rules are obeyable, persons intent on mayhem have not been promised or do not simultaneously seek the return of a cache of arms. Sororal obligations such as those of Antigone do not conflict with obligations to preserve peace, and so on. Agents may still fail to fulfill obligations.

Consider, for example, a silly two-person card game. (This is the partial analogue of a two-person dilemma. One can contrive silly games of solitaire for the one-person dilemma.) In the two person game the deck is shuffled and divided equally, face down between two players. Players turn up top cards on each play until the cards are played out. Two rules are in force: black cards trump red cards, and high cards (ace high) trump lower-valued cards without attention to color. Where no rule applies, e.g., two red deuces, there is indifference and the players proceed. We could define the winner as the player with the largest number of tricks when the cards are played out. There is an inclination to call such a set of rules inconsistent, for suppose the pair turned up is a red ace and a black deuce; who trumps? This is not a case of rule indifference as in a pair of red deuces. Rather, two rules apply, and both cannot be satisfied. But, on the definition here proposed, the rules are consistent in that there are possible circumstances where, in the course of playing the game, the dilemma would not arise and the game would proceed to a conclusion. It is possible that the cards be so distributed that, when a black card is paired with a red card, the black card happens to be of equal value. Of course, with shuffling, the likelihood of dilemma-free circumstances is small. But we could have invented a similar game where the likelihood of proceeding to a conclusion without dilemma is greater. Indeed, a game might be so complex that the likelihood of its being dilemmatic under any circumstances is very small and may not even be known to the players.[7] On the proposed definition, rules are consistent if there are possible circumstances in which no conflict will emerge. By extension, a set of rules is inconsistent if there are *no* circumstances, no possible world, in which all the rules are satisfiable.[8]

A pair of offending rules that generates inconsistency as *here* defined provides *no* guide to action under any circumstance. Choices are thwarted whatever the contingencies. Well, a critic might say, you have made a trivial logical point. What pragmatic difference is there

7. There is a question whether, given such rules, the "game" is properly described as a game. Wittgenstein says, "Let us suppose that the game (which I [Wittgenstein] have invented) is such that whoever begins can always win by a particular simple trick. But this has not been realized—so it is a game. Now someone draws our attention to it—and it stops being a game." *Remarks on the Foundation of Mathematics,* ed. G. H. von Wright et al., trans. G. E. M. Anscombe (Oxford: Blackwell, 1956), II, 78, p. 100e. Wittgenstein is pointing out the canon of a game that requires that both players have some opportunity to win. The canon that rules out dilemmatic rules is that the game must be playable to a conclusion. (I am beholden to Robert Fogelin for reminding me of this quotation.)

8. Bernard Williams, in *Problems of the Self* (New York: Cambridge University Press, 1977), chapters 11 and 12, also recognizes that the source of some apparent "inconsistencies" in imperatives and rules is to be located in the contingency of their simultaneous inapplicability on the given occasion.

between the inconsistent set of rules and a set, like those of the game described above, where there is a likelihood of irresolvable dilemma? A code is, after all, supposed to guide action. If it allows for conflicts without resolution, if it tells us in some circumstances that we ought to do *x* and we ought to do *y* even though *x* and *y* are incompatible in those circumstances, that is tantamount to telling us that we ought to do *x* and we ought to refrain from doing *x* and similarly for *y*. The code has failed us as a guide. If it is not inconsistent, then it is surely deficient, and, like the dilemma-provoking game, in need of repair.

But the logical point is not trivial, for there are crucial disanalogies between games and the conduct of our lives. It is part of the canon of the family of games of chance like the game described, that the cards must be shuffled. The distribution of the cards must be "left to chance." To stack the deck, like loading the dice, is to cheat. But, presumably, the moral principles we subscribe to are, whatever their justification, not justified merely in terms of some canon for games. Granted, they must be guides to action and hence not totally defeasible. But consistency in our sense is surely only a necessary but not a sufficient condition for a set of moral rules. Presumably, moral principles have some ground; we adopt principles when we have reasons to believe that they serve to guide us in right action. Our interest is not merely in having a playable game whatever the accidental circumstances, but in doing the right thing to the extent that it is possible. We want to maximize the likelihood that in all circumstances we can act in accordance with each of our rules. To that end, our alternative as moral agents, individually and collectively, as contrasted with the card-game players, is to try to stack the deck so that dilemmas do not arise.

Given the complexity of our lives and the imperfection of our knowledge, the occasions of dilemma cannot always be foreseen or predicted. In playing games, when we are faced with a conflict of rules we abandon the game or invent new playable rules; dissimilarly, in the conduct of our lives we do not abandon action, and there may be no justification for making new rules to fit. We proceed with choices as best we can. Priority rules and the like assist us in those choices and in making the best of predicaments. But, if we do make the best of a predicament, and make a choice, to claim that one of the conflicting obligations has thereby been retroactively erased is to claim that it would be mistaken to feel guilt or remorse about having failed to act according to that obligation. So the agent would be said to believe falsely that he is guilty, since his obligation was vitiated and his feelings are inappropriate. But that is false to the facts. Even where priorities are clear and overriding and even though the burden of guilt may be appropriately small, explanations and excuses are in order. But in such tragic cases as that described by Jean-Paul Sartre,[9] where the choice to be made by the agent is between not abandoning a wholly dependent mother and becoming a freedom fighter, it is inadequate to insist that feelings of guilt about the rejected alternative are mistaken and that assumption of guilt is inappropriate. Nor is it puritanical zeal that insists on the reality of dilemmas and the appropriateness of the attendant feelings, for dilemmas, when they occur, are data of a kind. They are to be taken into account in the future conduct of our lives. If we are to avoid dilemmas we must be motivated to do so. In the absence of associated feelings, motivation to stack the deck, to arrange our lives and institutions so as to minimize or avoid dilemma, is tempered or blunted.

Consider, for example, the controversies surrounding nonspontaneous abortion. Philosophers are often criticized for inventing bizarre examples and counterexamples to make a philosophical point. But no contrived example can equal the complexity and the puzzles generated by the actual circumstances of fetal conception, parturition, and ultimate birth of a human being. We have an organism, internal to and parasitic upon a human being, hidden from view but relentlessly developing into a human being, which at some stage of development can live, with nurture, outside its host. There are arguments that recognize competing claims: the right to life of the fetus (at some stage) versus the right of someone to determine what happens to her body. Arguments that justify choosing the mother over the fetus (or vice versa) where their survival is in competition. Arguments in which fetuses that are defective are balanced against the welfare of others. Arguments in which the claims to survival of others will be said to override survival of the fetus under conditions of great scarcity. There are even arguments that deny *prima facie* conflicts altogether on some metaphysical grounds, such as that the fetus is not a human being or a person until quickening, or until it has recognizable human features,

---

9. Sartre in "Existentialism Is a Humanism" (Paris: Nagel, 1946) describes a case where a student is faced with a decision between joining the Free French forces and remaining with his mother. He is her only surviving son and her only consolation. Sartre's advice was that "no rule of general morality can show you what you ought to do." His claim is that in such circumstances "nothing remains but to trust our instincts." But what is "trust" here? Does our action reveal to us that we subscribe to a priority principle or that in the absence of some resolving principles we may just as well follow our inclination? In any case to describe our feelings about the rejected alternative as "regret" seems inadequate. See Walter Kaufmann, ed., *Existentialism from Dostoevsky to Sartre* (New York: Meridian, 1956), pp. 295–98.

or until its life can be sustained external to its host, or until birth, or until after birth when it has interacted with other persons. Various combinations of such arguments are proposed in which the resolution of a dilemma is seen as more uncertain, the more proximate the fetus is to whatever is defined as being human or being a person. What all the arguments seem to share is the assumption that there is, despite uncertainty, a resolution without residue; that there is a correct set of metaphysical claims, principles, and priority rankings of principles that will justify the choice. Then, given the belief that one choice is justified, attribution of guilt relative to the overridden alternative is seen as inappropriate, and feelings of guilt or pangs of conscience are viewed as, at best, sentimental. But as one tries to unravel the tangle of arguments, it is clear that to insist that there is in every case a solution without residue is false to the moral facts.

John Rawls,[10] in his analysis of moral sentiments, says that it is an essential characteristic of a moral feeling that an agent, in explaining the feeling, "invokes a moral concept and its associated principle. His (the agent's) account of his feelings makes reference to an acknowledged right or wrong." Where those ingredients are absent, as, for example, in the case of someone of stern religious background who claims to feel guilty when attending the theater although he no longer believes it is wrong, Rawls wants to say that such a person has certain sensations of uneasiness and the like that resemble those he has when he feels moral guilt, but, since he is not apologetic for his behavior, does not resolve to absent himself from the theater, does not agree that negative sanctions are deserved, he experiences not a feeling of guilt but only something like it. Indeed, it is the feeling that needs to be explained; it is not the action that needs to be excused, for says Rawls, in his discussion of moral feelings and sentiments, "When plagued by feelings of guilt . . . a person wishes to act properly in the future and strives to modify his conduct accordingly. He is inclined to admit what he has done, to acknowledge and accept reproofs and penalties." Guilt *qua* feeling is here defined not only in

10. *A Theory of Justice* (Cambridge: Harvard University Press, 1971), pp. 481–83. Rawls's claim is that such sensations, to be properly describable as "guilt feelings" and not something resembling such feelings, must occur in the broader context of beliefs, strivings, acknowledgments, and readiness to accept outcomes, and cannot be detached from that context. He rejects the possibility that there are such "pure" sensations that can occur independent of the broader context. This is partially, perhaps, an empirical claim about identifying sameness of feeling. The theatergoer might claim that he does feel guilty because he has the same feeling he has when he acknowledges that he is guilty, that what remains is to give an account of when such feelings of guilt are justified. Still, Rawls's analysis seems to me to be a better account.

terms of sensations but also in terms of the agent's disposition to acknowledge, to have wishes and make resolutions about future actions, to accept certain outcomes, and the like. Where an agent acknowledges conflicting obligations, unlike the theatergoer, who acknowledges no such obligation, there is sufficient overlap with dilemma-free cases of moral failure to warrant describing the associated feelings where present as guilt, and where absent as nevertheless appropriate to an agent with moral sensibility. Granted that, unlike agents who fail to meet their obligations *simpliciter,* the agent who was confronted with a dilemma may finally act on the best available reasons. Still, with respect to the rejected alternative he acknowledges a wrong in that he recognizes that it was within his power to do otherwise. He may be apologetic and inclined to explain and make excuses. He may sometimes be inclined to accept external reproofs and penalties. Not perhaps those that would be a consequence of a simple failure to meet an obligation but rather like cases in which mitigating circumstances evoke a lesser penalty.[11]

Even if, as Rawls supposes, or hopes (but as seems to me most unlikely), a complete set of rules and priorities were possible that on rational grounds would provide a basis for choosing among competing claims in all cases of moral conflict that actually arise, it is incorrect to suppose that the feeling evoked on such occasions, if it is evoked, only resembles guilt, and that it is inappropriate on such occasions to so characterize those sentiments. *Legal* ascriptions of guilt require sanctions beyond the pangs of conscience and self-imposed reproofs. In the absence of clear external sanctions, legal guilt is normally not ascribable. But that is one of the many distinctions between the legal and the moral.

Most important, an agent in a predicament of conflict will also "wish to act properly in the future and strive to modify his actions accordingly." That should reasonably include striving to arrange his own life and encourage social arrangements that would prevent, to the extent that it is possible, future conflicts from arising. To deny wholly the appropriateness or correctness of ascriptions of guilt is to weaken the impulse to make such arrangements.[12]

---

11. To insist that ''regret'' is appropriate rather than ''guilt'' or ''remorse'' is false to the facts. It seems inappropriate, for example, to describe as ''regret'' the common feelings of guilt that women have in cases of abortion even where they believe that there was moral justification in such an undertaking.

12. Bernard Williams (''Politics and Moral Character,'' in *Public and Private Morality,* ed. Stuart Hampshire [New York: Cambridge University Press, 1978], pp. 54–74) discusses the question of moral conflict in the context of politics and the predicament of ''dirty hands.'' He argues that, where moral ends of politics justify someone in public life

III

I have argued that the consistency of a set of moral rules, even in the absence of a complete set of priority rules, is not incompatible with the reality of moral dilemmas. It would appear, however, that at least some versions of the principle " 'ought' implies 'can' " are being denied, for dilemmas are circumstances where, for a pair of obligations, if one is satisfied then the other cannot be. There is, of course, a range of interpretations of the precept resulting from the various interpretations of 'ought', 'can', and 'implies'. Some philosophers who recognize the reality of dilemmas have rejected the precept that " 'ought' implies 'can' "; some have accepted it.[13] If we interpret the 'can' of the precept as "having the ability in this world to bring about," then, as indicated above, in a moral dilemma, 'ought' *does* imply 'can' for *each* of the conflicting obligations, *before* either one is met. And after an agent has chosen one of the alternatives, there is still something that he ought to have done and could have done and that he did not do. 'Can', like 'possible', designates a modality that cannot always be factored out of a conjunction. Just as 'possible P and possible Q' does not imply 'possible both P and Q', so 'A can do x and A can do y' does not imply 'A can do both x and y'. If the precept " 'ought' implies 'can' " is to be preserved, it must also be maintained that 'ought' designates a modality that cannot be factored out of a conjunction. From 'A ought to do x' and 'A ought to do y' it does not follow that 'A ought to do x and y'. Such a claim is of course a departure from familiar systems of deontic logic.

The analysis of consistency and dilemmas advanced in this paper suggests a second-order principle that relates 'ought' and 'can' and that provides a plausible gloss of the Kantian principle "Act so that thou canst will thy maxim to become a universal law of nature." As Kant understood laws of nature, they are, taken together, universally

---

lying, or misleading, or using others, "the moral disagreeableness of these acts is not merely canceled." In particular, we would not want, as our politicians, those "practical politicians" for whom the sense of disagreeableness does not even arise.

13. For example, John Lemmon, in "Moral Dilemmas," *Philosophical Review,* LXXI, 2 (April 1962): 137–158, p. 150, rejects the principle that 'ought' implies 'can'. Van Fraassen, "Values and the Heart's Command," pp. 12–13, accepts it, as does Bernard Williams seemingly in *Problems of the Self,* pp. 179–184. Van Fraassen and Williams see that such acceptance requires modification of the principles of factoring for the deontic 'ought.' There are other received principles of deontic logic that will have to be rejected, but they will be discussed in a subsequent paper. It should also be noted that, in "Ethical Consistency" and "Consistency and Realism" in *Problems of the Self,* Williams also articulates the contingent source of dilemmas and argues for their "reality" but he sees them as an argument against moral realism.

and jointly applicable in all particular circumstances. It is such a second-order principle that has been violated when we knowingly make conflicting promises. It is such a second-order principle that has, for example, been violated when someone knowingly and avoidably conducts herself in such a way that she is confronted with a choice between the life of a fetus, the right to determine what happens to one's body, and benefits to others. To will maxims to become universal laws we must will the means, and among those means are the conditions for their compatibility. One ought to act in such a way that, if one ought to do x and one ought to do y, then one can do both x and y. But the second-order principle is regulative. This second-order 'ought' does *not* imply 'can'.[14] There is no reason to suppose, this being the actual world, that we can, individually or collectively, however holy our wills or rational our strategies, succeed in foreseeing and wholly avoiding such conflict. It is not merely failure of will, or failure of reason, that thwarts moral maxims from becoming universal laws. It is the contingencies of this world.

IV

Where does that leave us? I have argued that all dilemmas are real in a sense I hope has been made explicit. Also that there is no reason to suppose on considerations of consistency that there *must* be principles that, on moral grounds, will provide a sufficient ordering for deciding all cases. But, it may be argued, when confronted with what are *apparently* symmetrical choices undecidable on moral grounds, agents do, finally, choose. That is sometimes understood as a way in which, given goodwill, an agent makes explicit the rules under which he acts. It is the way an agent discovers a priority principle under which he orders his actions. I should like to question that claim.

A frequently quoted remark of E. M. Forster[15] is "If I had to choose between betraying my country and betraying my friend, I hope I should have the courage to betray my country." One could of course read that as if Forster had made manifest some priority rule: that certain obligations to friends override obligations to nation. But consider a remark of A. B. Worster, "If I had to choose between betraying my country and betraying my friend, I hope I should have the courage

14. See note 13 above. The reader is reminded that, on the present analysis, 'ought' is indexical in the sense that applications of principles on given occasions project into the future. They concern bringing something about.
15. *Two Cheers for Democracy* (London: E. Arnold, 1939).

to betray my friend.'' Both recognize a dilemma, and one can read Worster as subscribing to a different priority rule and, to that extent, a different set of rules from Forster's. But is that the only alternative? Suppose Forster had said that, morally, Worster's position is as valid as his own. That there was no moral reason for generalizing his own choice to all. That there was disagreement between them not about moral principles but rather about the kind of persons they wished to be and the kind of lives they wished to lead. Forster may not want Worster for a friend; a certain possibility of intimacy may be closed to them that perhaps Forster requires in a friend. Worster may see in Forster a sensibility that he does not admire. But there is no reason to suppose that such appraisals are or must be moral appraisals. Not all questions of value are moral questions, and it may be that not all moral dilemmas are resolvable by principles for which *moral* justification can be given.

10. *Rationality and Believing the Impossible*

The present paper appeared in the *Journal of Philosophy,* LXXX, 6 (June 1983): 321–338. It was written during a memorable 1982 stay at U.C.L.A. Variants were presented to several audiences whose comments were valuable and whose patience with my appeals to unshared intuitions, was superogatory. Given the absence of consensus on how one is to understand claims to beliefs in impossibilities, I modified the position taken here in a later overlapping essay, "Some Revisionary Proposals about Beliefs and Believing," also included in this volume. ■

There are some familiar shared views about the use of 'knows that' whereby if someone, call him an epistemological agent, claims to know that *p* and if *'p'* is false, i.e., *'p'* does not describe an actual state of affairs, then the agent is mistaken in his knowledge claim. He does not say, on discovering the falsity of *'p'*, that once he knew that *p* and now he knows that not-*p*, that false knowledge has been replaced by true knowledge. He will say, rather, that he was mistaken in the first instance in claiming knowledge. It was perhaps more correctly a belief he had that *p*. Those who insist otherwise are seen as conceptually confused. The truth of a sentence that makes a knowledge claim is, on this widely shared view, dependent on the way things are, however strong the justifications advanced and the behavioral or psychological evidence that the agent believed that *p*. It is also part of this common view that an agent's *belief* that *p* is independent of the truth of *'p'*: *'p'* can describe a nonactual state of affairs, and yet the agent can have that belief. On discovering the falsity of *'p'*, the agent is *not* conceptually confused when he says that once he believed that but, on discovering that not-*p*, he no longer believes that *p*. He does not say that his claiming to believe was mistaken in the first instance.

Given the view that, unlike knowing, believing is a state that can obtain independent of the way the world is, it is often taken as a consequence that believing must be a purely mental state and that an account of belief can be given in psychological terms. But, given that 'believes' is a verb that takes an object, there arises the question of the nature of the object. It has been customary to call such objects "propositions," and, as we know, their nature has been a considerable source of bafflement. If one takes those objects, propositions, as themselves mental entities, then the psychological purity of believing can be preserved—a question to which I will return.[1] At this juncture it

---

1. A difficulty with discussions of epistemological attitudes is the equivocation engendered by shuttling between describing beliefs as psychological states or features of psychological states and describing beliefs as objects of believing. Recently Daniel Dennett, after hearing an earlier version of this paper, sent me a copy of his newly published, rich, and interesting monograph "Beyond Belief" in *Thought and Object: Essays on Intentionality*, ed. Andrew Woodfield (New York: Oxford University Press, 1982). Much of what he says in criticism of some familiar positions about belief supports views presented in this paper and in my "A Proposed Solution to a Puzzle about Belief," in *Midwest Studies in Philosophy: Foundations of Analytic Philosophy*, vol. VI, ed. P. French et al. (1981), pp. 501–510. But even in "Beyond Belief," pp. 2–3, the incautious use of 'belief' for the state as well as the object is an impediment to a more clearly articulated position. Dennett says, for example, "If we still want to talk about beliefs (pending such a discovery) we must have some way of picking them out, distinguishing them from each other. If beliefs are real—that is—real psychological states of people, there must be indefinitely many ways

is enough to say that I will urge in what follows that a more adequate account of belief will take believing to be a relation between an agent and a possible state of affairs, whereas knowing is a relation between an agent and, more restrictively, an actual state of affairs.

As noted above, on the common view, the falsehood of '*p*' does not rule out believing that *p*. However, where '*p*' is necessarily false, false under any circumstances, there has been some divergence as to whether an agent can properly be said to believe or to have believed that *p*, or whether on the contrary such a belief claim is mistaken. Views of those who support the latter alternative are largely discounted, for evidence seems to support the claim that the necessary falsehood of '*p*' does not preclude believing that *p*. Mathematical conjectures, it is argued, are, if false, necessarily so; yet some mathematical conjectures, purportedly believed by competent mathematicians who do not suffer from conceptual confusion, have subsequently been demonstrated to be false. Also, if one accepts (as I do) the principle that logically irreducible identity sentences (i.e., sentences where the names flanking the identity sign are proper names) are, if true, necessarily so, then there are basic, noncomplex examples that support the claim that we can believe the impossible, or, as I prefer to put it, that we can enter into the belief relation with an impossible state of affairs, for what is described by a false identity sentence is an impossibility, metaphysically speaking.

But there *is* a weaker constraint on believing the impossible, shared by many who hold the dominant view. The constraint is not the strong one of supposing that an epistemological agent is a perfect logician, a perfect deducer, or a perfect appraiser of evidence; referring to ordinary reasoners, Wilfrid Hodges[2] remarks in his *Logic:* "It is simply

---

of referring to them." Here Dennett takes beliefs to be psychological states. But shortly thereafter he goes on to say, "The privileged way of referring to beliefs, what we usually mean and are taken to mean in the absence of special provisions or contextual cues, is the proposition believed." Here it is presumably the object of the psychological state that is under discussion.

It may be a consequence of such incautious use that, in the absence of psychological objects that can be plausibly *identified* with beliefs, i.e., the objects of believing, we had better abandon the search for a theory about belief—a position that seems to be advanced by Dennett.

2. New York: Penguin Books, 1977, p. 15. Hodges does not present us with substantial arguments in support of his view; he asks us to try to convince ourselves, where we know that *p*, that we might also believe that not-*p*. That effort, he suggests, is supposed to persuade us of the impossibility of believing both. It is fair to understand his position as suggesting that, if an agent knows or believes that *p* and *q* are incompatible, he would be conceptually confused if he also claimed to believe both *p* and *q*. But, it may be asked, why the stress on inconsistency? Is it possible to believe *p* where we know that *p* is false? See also Gilbert Harman, "Induction," in *Introduction, Acceptance, and Rational Belief,* ed. Marshall Swain (Boston: Reidel, 1970), pp. 98–99.

impossible to believe . . . two things which you *know* are inconsistent with each other. It seems we are obliged to believe only what we *think* is consistent.'' Accordingly, a necessary condition for an agent's having a belief that *p,* on this widely accepted modified view, is that the agent not know or, more weakly, not believe that *p* is impossible.

Those who have pondered the question of intensional contexts will see that even this modified constraint is far from clear. But it does appear to be a weaker constraint on the concept of belief than denying *simpliciter* that an agent can believe the impossible. On this modified view the White Queen can be seen as conceptually confused, at least on a *de re* reading, when, in response to Alice's denial that one can believe the impossible, the White Queen says, "Why sometimes I've believed as many as six impossible things before breakfast." [3] But the modified constraint on believing the impossible is concerned with an agent's knowing or believing that what he believes is impossible; it suggests that, in the absence of such second-order epistemic attitudes, impossibility does not preclude belief.

But there is a still stronger view, which I would like to defend: Alice's view, that we cannot believe an impossibility. I would like to defend it not merely as an analytical exercise, but because it has consequences for theories about epistemological attitudes and theories of rationality. It seems to me that there is greater symmetry between knowing and believing than is allowed by prevailing views, even where minimally modified as above, with a second-order constraint. The position I propose is that, just as knowing that *p* relates an agent to an actual state (or states) of affairs, otherwise the knowledge claim [the claim that one knows] is mistaken, so there is an important and neglected sense of 'believes' such that believing *p* relates an agent to a possible state (or states) of affairs, otherwise the belief claim [the claim that one believes] is mistaken despite the apparent evidence to the contrary.

Before proceeding to an account that accommodates seemingly contrary evidence, we should first note that this stronger constraint on belief is not wholly idiosyncratic. It can be found with some frequency in the literature. Among the examples is one I find particularly perspicuous. In *Principles of Human Knowledge*[4] George Berkeley

---

3. Lewis Carroll, in *Through the Looking Glass;* see *The Annotated Alice,* ed. Martin Gardner (New York: New American Library, 1960), p. 251. On a *de dicto* reading the White Queen might be claiming that, of her large number of beliefs, she is prepared to say that six are necessarily false, without specifying which.
4. Ed. C. M. Turbayne (Indianapolis: Bobbs-Merrill, 1970), p. 273. Berkeley shares the received view in supposing that any statement or conjunction of statements that de-

claims to have *demonstrated* that the existence of mindless matter is impossible and that it involves a "contradiction." He then goes on to say (paragraph 54):

> Strictly speaking, to believe that which involves a contradiction . . . is impossible. . . . In one sense, indeed, men may be said to believe that matter exists; that is, they act *as if* the immediate cause of their sensations, which affects them every moment . . . were some senseless unthinking being. But that they . . . should form thereof a settled speculative opinion is what I am unable to conceive.

and, finally, Berkeley concludes,

> This is not the only instance wherein men impose upon themselves, by imagining they believe those propositions they have often heard, though at bottom they (those propositions) have no meaning in them.

Since Berkeley speaks of *hearing* propositions, we may suppose him to be saying that sentences that describe impossible states of affairs are meaningless. My proposal is not so strong. I want to propose that, where a state of affairs is impossible, there is a sense of "belief" such that an agent is mistaken if he claims that he is in the belief relation to that state of affairs. In such cases, as Berkeley remarks, agents behave in some respects *as if* they were in that relation, or, elliptically, *as if* they had that belief.

It is helpful at this juncture to consider how the received view explains apparent belief in the impossible. The explanation is, on the surface, simple, but only because the view takes, as the objects of belief, linguistic or quasi-linguistic entities, usually called "propositions," which, in many of their features, are or mimic interpreted sentences of a language.[5] Propositions are said to have properties like

---

scribes an impossibility must be "a contradiction." But, as is clear from the account that follows, there are statements that describe necessary and impossible states of affairs but that are not, on the usual account of such properties, *logically* valid or contradictory. It has become common practice to say of such statements—for example, attribution of kind properties or identity statements—that they describe metaphysical necessities or metaphysical impossibilities.

More recently, one can locate in Wittgenstein's *Tractatus* and among some positivists the view that impossible propositions are meaningless, i.e., they have no cognitive content. Where '*p*' is such a proposition 'Pierre believes that *p*' is taken to be false or meaningless.

5. There are, of course, and especially recently, uses of 'proposition' that do not wholly identify propositions with linguistic or quasi-linguistic entities. I am thinking, for example, of (1) some versions of possible-world semantics that take propositions to be sets of worlds that share certain features, as in R. Stalnaker, "Indexical Belief," *Synthese*,

truth and validity. Sets of them are consistent or inconsistent. They can be contradictory. They can enter into the consequence relation. On this common view propositions are accepted and asserted. Some authors[6] even speak of *assenting* to them, as well as believing them or knowing them.

The identification of beliefs, i.e., the objects of believing, with sentence-like objects has some familiar consequences. The believer, the agent, *has* those beliefs. If they are quasi-linguistic entities, how does the agent *have* them? For Frege, there is that common stock of thoughts, those abstract language-like entities, seemingly outside the mind (despite their being called "thoughts") toward which the agent has an attitude. But then how does the agent *have* them? Less mysteriously, on some contemporary views the agent is seen as having an internal register of sentences or "mental representations" of sentences associated with either "yes" or "no" responses on appropriate cues. The agent then is said to believe that *p* just in case the program for a correlated mental representation "sentence" elicits a "yes" response from the register. This is the thrust, oversimplified of course, of an analysis of belief that takes the objects of belief to be linguistic or quasi-linguistic entities. The computer model, or the model of mental representation such as that of Jerry Fodor,[7] has at least the advantage over Frege of trying to make more explicit the linguistic nature of propositions, and places those linguistic objects squarely in the mind. Fodor tells us, without flinching, that attitudes toward prop-

---

XLIX, 1 (1981): 129–151; (2) David Kaplan's view of propositions as "contents" ("Locke Lectures," unpublished); (3) J. Perry and J. Barwise's "situation semantics," which takes propositions to be courses of events; see "Situations and Attitudes," *Journal of Philosophy*, LXXVIII, 11 (November 1981): 668–691. In 2 and 3 we come closer to the Russellian view of propositions, also reflected in R. Chisholm's view of "propositions" as states of affairs, "Events and Propositions," *Noûs*, IV, 1 (1970): 15–24. See also A. Plantinga, *The Nature of Necessity* (New York: Oxford University Press, 1974).

Given the widespread and persistent use of 'proposition' as a linguistic entity, a use that is justified by etymology, we would do well to abandon employment of the term for designating nonlinguistic entities.

6. See, for example, Daniel Dennett, *Brainstorms* (Montgomery, Vt.: Bradford Books, 1978), chap. 1, pp. 3–22, especially p. 21, where he says, "One might have a theory about an individual's neurology that permitted one to 'read off' or predict the propositions to which he would assent, but whether one's theory had uncovered his *beliefs*, or merely a set of assent-inducers, would depend on how consistent, reasonable, true we found the set of propositions." In fairness to Dennett, he has, since writing chapter 1, considerably modified the extent to which he subscribes to the received view. Indeed, in chapter 16 of *Brainstorms*, Dennett draws upon a distinction between assenting and believing made originally by Ronald de Sousa in "How to Give a Piece of Your Mind," *Review of Metaphysics*, XXV (1971): 52–79. The distinction as drawn by Dennett has some points of overlap with the distinction between assenting and believing of the present paper. And, if one reads carefully, one finds that it is similarly motivated.

7. *The Language of Thought* (New York: Crowell, 1975).

ositions are in fact "attitudes" toward formulas, which are internally codified and which represent the external sentences of a given language.

It is of historical interest that F. P. Ramsey,[8] in "Facts and Propositions," expanding on some early views of Russell, initially proposed a theory in which believing relates "mental factors" (including mental representations) to "objective factors" in the world. Believing, therefore, in the first instance is not viewed as confined to relationships to linguistic or quasi-linguistic entities. But Ramsey's theory, finally, is not significantly divergent from language-centered theories of belief, for he goes on to say that the nature of mental factors will

> depend on the sense in which we are using the ambiguous term 'belief': it is for instance possible to say that a chicken believes a certain sort of caterpillar to be poisonous, and mean by that merely that it abstains from eating such caterpillars on account of unpleasant experiences connected with them. . . . But without wishing to depreciate the importance of this kind of belief . . . I prefer to deal with those beliefs which are expressed in words, or possibly images or other symbols, consciously asserted or denied. . . . The mental factors of such a belief I take to be words spoken aloud . . . or to oneself or merely imagined, connected together and accompanied by a feeling or feelings of belief.

Having said that, Ramsey focuses on attitudes toward the *sentences* that are vehicles for "expressing" or "picturing" the objective factor in belief. As a consequence, 'assenting to *p*', 'asserting that *p*' become for him alternative usages. It is part of the thrust of this paper to claim that Ramsey was mistaken in viewing 'belief' as ambiguous. Rather, I will urge that the disposition to assent to sentences that are thought to "express" what one believes, in language, is one of the many kinds of dispositions to behavior that are evidence for an agent's believing. Of course such linguistic dispositions and linguistic behavior are closed to the chicken.

In what follows I want to argue that to fix on linguistic or quasi-linguistic entities as the objects of belief cannot be supported within an adequate account of epistemological attitudes. However, *if* one does adopt that view, then accounting for belief in the impossible *appears* straightforward. On the assumption that the agent is not conceptually

8. In *The Foundations of Mathematics* (New York: Humanities Press, 1950), pp. 138–155; the quotation is on p. 144. Ramsey here revives and considers a former view of Russell's in which believing is seen as relating mental factors in an agent to an objective factor.

confused,[9] an agent can come to believe an impossibility if he is less than ideally rational, narrowly defined. An ideally rational agent, narrowly defined, is one who would believe all the logical consequences of his beliefs and would not believe a conjunction of propositions that make up an inconsistent set. But agents are less than ideally rational and hence unknowingly (given the minimal constraint) may come to believe an impossibility.[10]

Nonetheless, appearances to the contrary, departure from perfect rationality as above characterized still does not fully explain, on certain plausible assumptions of the received view, how an agent can come to have such a belief. Consider an agent at some moment in time who assents to each of a certain set of sentences. Suppose him to be sincere, reflective, not conceptually confused although not omniscient. Suppose him also to be logically perfect in the narrow sense that (1) he does not and would not assent to a conjunction of sentences that make up an inconsistent set; (2) he assents or would assent to the logical consequences of the sentences to which he assents. Indeed, on this narrow view of "rationality" he is perfectly rational. Yet it seems that such an agent can also find himself in the apparent predicament of believing an impossibility. Indeed, it seems that it is his *very rationality* that lands him in the predicament. Saul Kripke[11] gives us examples: instances where, on generally accepted principles relating assent to belief, a logically perfect agent would seem to believe an impossibility. Kripke takes this puzzle to be deeply puzzling. I take it as an argument against the received view that accepts as noncontroversial that certain entailment relations must hold between a particular linguistic activity, such as assenting, and believing.[12]

It is the thrust of this paper that, once we question whether certain entailment relations must hold between assenting and believing, the way is open not merely to a solution of such puzzles but to a more general theory about epistemological attitudes and rational belief. But

9. Roughly, a conceptual confusion is one that could be resolved by resorting to a standard lexicon.

10. Rationality, as usually understood, does not require omniscience. On the received view it does require that beliefs (which are "propositions") make up a consistent set. Also, it requires that belief be closed under logical consequence. Given the difficulties posed for a theory of belief for the less than ideally rational agent, Jaakko Hintikka in *Knowledge and Belief* (Ithaca: Cornell University Press, 1962) develops a theory for the ideally rational agent. Dennett, despairing of arriving at a theory for the less than ideally rational agent, urges that we should "cease talking about belief and descend to the design stance or physical stance for one's account," *Brainstorms*, p. 22.

11. "A Puzzle about Belief" in *Meaning and Use*, ed. Avishai Margalit (Dordrecht: Reidel, 1979), pp. 234–283.

12. See my "A Proposed Solution to a Puzzle about Belief."

first it will be necessary to review briefly Kripke's account of the puzzle.

Kripke begins with the assumption that the language users under discussion are nonomniscient normal speakers, not conceptually confused, and sincere and reflective. It is also assumed that the sentences in (1) below do not contain demonstratives, indexicals, and the like.[13] Given such assumptions, it is taken as a noncontroversial principle (the disquotation principle) that

(1)   If a speaker of a language L assents to '*p*' and '*p*' is a sentence of L, then he believes that *p*.

The conditional in (1) is an entailment. Given this disquotation principle, examples are cited where, in conjunction with other plausible claims, a logically perfect rational agent can come to believe an impossibility. Kripke's example is a bilingual case that also requires a noncontroversial principle of translation.

(2)   If a sentence of one language expresses a truth in that language, then any translation of it into another language also expresses a truth in that other language.

The case is as follows. Pierre, a native, initially monolingual Frenchman, has heard or read about London's being pretty. He has perhaps seen pictures. (His name for London is 'Londres') So, he assents to the sentence

(3)   Londres est jolie.

He emigrates to England, takes up residence in London, learns English by exposure, and, given his observation of his surroundings, assents to

(4)   London is not pretty.

Given that 'London' and 'Londres' have the same fixed semantic value, it follows from the disquotation principle (1) and the translation principle (2) that Pierre believes that London is pretty and he believes that London is not pretty. He has not made any logical blunders, he is not a poor appraiser of evidence, he has not replaced one belief about London by another. Of course it does not follow from the disquotation principle (1) that Pierre assents to 'London is pretty'.

13. Consider for example *my* report about John under the disquotation principle: "If John assents to (his utterance) 'I was born in California' then John believes that I was born in California."

If it is also taken as plausible and noncontroversial in the "logic" of belief (as I will later deny) that belief factors out of a conjunction, that if believing *p* and believing *q* entail believing *p* and *q*, then Pierre believes an impossibility.

If it is claimed that the puzzle shows the weakness of the theory of direct reference of proper names, that is mistaken. Fregean theories as ordinarily understood fare no better.[14]

Nor can the predicament be explained as a conceptual confusion to be resolved by a standard lexicon. Proper names once given are fixed in their semantic values by the language viewed as an historical institution evolving over time. But it is not arbitrary what, *once fixed,* those values are. It may not be known to or believed by an entire community of speakers of a language in a given historical time slice, that two proper names have the same semantic value, but they may have the same value nevertheless. An encyclopedia would simply be mistaken if it claimed the contrary. On the assumption that the Hesperus-Phosphorus story is accurate, an encyclopedia, published before certain astronomical discoveries, might claim that Hesperus was different from Phosphorus and be endorsed by universal assent to the sentence 'Hesperus is different from Phosphorus', but that claim would be mistaken. In fact, as I shall urge, it cannot be believed that Hesperus is different from Phosphorus, appearances or encyclopedias to the contrary.

I should like to begin addressing the puzzle by proposing a modification of the disquotation principle. The modification may seem ad hoc, but what is of interest is not the seemingly arbitrary modification *per se* along Berkeleyan lines but the underlying motivation, which is to escape the strong language-centricity of prevalent theories about epistemological attitudes and their objects and to allow for a better account of rationality. The modified disquotation principle is (retaining assumptions given for (1)) as follows:

(1′)   If a speaker of a language L assents to '*p*' and '*p*' is a sentence of L and '*p*' describes a possible state or states of affairs, i.e., *p* is possible, then that speaker believes that *p*.

I should first like to appeal to some intuitions in support of the modification. These intuitions are not universally shared, but neither are they altogether idiosyncratic.

14. See Kripke, "A Puzzle about Belief." Of course I have no disagreement with the theory of direct reference for proper names. In fact I proposed such a theory in 1961. See "Modalities and Intensional Languages," this volume.

Beliefs are supposed to guide a range of our actions, not just our speech acts, and someone who claims to believe that states of affairs obtain that are impossible may, in some of his actions, like assenting on given occasions to certain sentences, be acting in some respects as if those belief states obtained. Pierre may, for example, sincerely announce, ''I intend to move from ugly London to beautiful Londres.'' But my intuition about my own language use is that, in a case like Pierre's, once an impossibility had been *disclosed,* I would say that I had only *claimed* to believe that London was different from Londres, for to have believed that Londres and London were not the same would be tantamount to believing that something was not the same as itself, and surely I could never believe *that.* So my belief *claim* (my claiming to believe) was mistaken, just as, on disclosure that a possible state of affairs does not *actually* obtain, I say that my knowledge claim (my claiming to know), if I have made one, was mistaken. Briefly, a proper object of believing is a possible state of affairs, and a proper object of knowing is an actual state of affairs.[15]

15. This ''intuition,'' shared by some and rejected by others, is supported indirectly by an experiment described by Donald Davidson in *Actions and Events* (New York: Oxford University Press, 1970), pp. 235–236. Davidson says, ''After spending several years testing variants of Ramsey's theory on human subjects, I tried the following experiment (with Merrill Carlsmith). Subjects made all possible pairwise choices within a small field of alternatives, and in a series of subsequent sessions, were offered the same set of options over and over. The alternatives were complex enough to mask the fact of repetition, so that subjects could not remember their previous choices, and pay-offs were deferred to the end of the experiment so that there was no normal learning or conditioning. The choices for each session and each subject were then examined for inconsistencies—cases where someone had chosen *a* over *b, b* over *c,* and *c* over *a.* It was found that as time went on, people became steadily more consistent; intransitivities were gradually eliminated; after six sessions, all subjects were close to being perfectly consistent . . . apparently, from the start there were underlying and consistent values which were better and better realized in choice. I found it impossible to construct a formal theory that could explain this, and gave up my career as an experimental psychologist.''

In Davidson's experiment it is as if the subjects who assented to sentences that are inconsistent declined to carry them over into belief. Such a subject, if my intuition is shared, would not say ''I once believed I preferred *a* to *b,* and *b* to *c,* and *c* to *a''* but now I don't. He would disclaim having had such a belief despite his acknowledged assents. A closer analogy to Pierre's case could be achieved if, *unknown* to Davidson's subject, the sentence '*a* = *d*' was true and he asserted 'I prefer *a* to *d*'.

Such tests of Ramseyan theories, where ''preferences'' are expressed in language rather than in the actual selection of *objects* in preference order, are susceptible to puzzles about assent and belief.

In Pierre's case his assent to a sentence describing an impossibility is not a logical lapse. It has its source in a lack of information. But on disclosure he also would, if he shared my intuition, disclaim having believed the impossibility.

An analogy with 'wanting' is helpful here in pressing the intuition. David Kaplan points out that an agent might, for example, have an aversion to mayonnaise. He says to the restaurant waiter that he wants the house dressing. But, on perceiving that the house dressing contains mayonnaise, the agent doesn't say that he formerly wanted the

In order to discuss the motivation and explanation of that intuition it is of primary importance to disengage particular acts on particular occasions, such as given acts of assenting, from believing—a separation rarely made, given the acceptance of the disquotation principle as noncontroversial. On the face of it, assenting on a given occasion is a particular speech act, which, on the face of it, believing is not. Even if one adopts the mental-representation view of belief, where actual assents are external correlates of internally registered assentings, these should be disengaged if one is to make clear the nature of their connection.

That connection, however, should not, even on the mental-representation view, be presupposed. If one denies that the mental-representation view gives us a plausible account of belief, as I do, the necessity for decoupling specific manifestations of believing, such as assenting on given occasions, *from* believing is even more crucial.

Both the unmodified and the modified disquotation principles apply only to language users and assert an essential connection for such users, between some speech acts, such as assenting, and believing. But there are good grounds for supposing that non-language-users, e.g., animals and preverbal children, have beliefs. To deny belief to the dog who accompanies his master to the store and waits quietly outside for his master to emerge is as anthropocentric as Descartes' denying pain to an animal undergoing a surgical procedure despite behavioral evidence to the contrary. Speech acts are only a part of the whole range of kinds of behavior that are manifestations of what we call "belief." Naturally, non-language-users will fail to have beliefs possible to language users—beliefs about language, for example. They will not have beliefs about describing and referring, validity, logical consequence, and the like. They will very likely be severely limited in the extent to which beliefs can be "communicated." They may have relatively meager beliefs in the absence of language that describes the objects of belief and without the use of a panoply of rules or procedures that relate such descriptions, such as deductive rules. They will not make belief claims or have second-order beliefs. But that is not to deny them beliefs altogether. The behavioral manifestations of belief just will not include speech acts.

Given the prevalence and entrenchment of the anthropocentric, or, more accurately, the linguacentric view of belief, i.e., that believing is a relationship open only to language users, it is perhaps worth be-

---

house dressing but no longer wants it. He says rather that he was mistaken in claiming he wanted it.

laboring the point. Denial of belief to non-language-users runs counter to a whole range of our experience. Choosing is supposed to be a manifestation of belief. Psychologists report on the behavior of pre-verbal children and animals by allowing them choices; allowing them, for example, to choose from a variety of foods in any amount as compared with disallowing choice of kind and amount. If in such an experiment with an animal who repeatedly selected kibbles from an array of options, the experimenter under similar conditions replaces kibbles by kibbles-facsimiles and the dog heads for them and takes a mouthful before rejecting them, it is not anthropomorphic for us to say that the dog was mistaken in believing that they were kibbles. He had a mistaken belief. He had a relationship to a possible, albeit nonactual, state of affairs—where real kibbles might have been in the dish. He cannot of course describe that possible state of affairs. Similarly, in the case of the dog waiting for his master to emerge from the store when in fact the master has gone out another door and abandoned the dog, it is not anthropomorphic to say that the dog had the mistaken belief that his master was in the store. If we do not require that the objects of belief be quasi-sentences—those ''propositions'' of the received view—we need not suppose that having language is a condition either for having beliefs or for having mistaken beliefs.

Yet consider Donald Davidson's[16] remarks: ''Someone cannot have a belief unless he understands the *possibility* of being mistaken, and this requires grasping the contrast between truth and error—true belief and false belief.'' He goes on: ''The notion of a true belief depends on the notion of a true utterance, and this in turn there cannot be without shared language.''

Now, it is likely that, for language users, there is an essential connection between the notions of truth and falsity as applied to utterances and being in the believing relation to actual and nonactual states of affairs—language users do have such notions as true and false taken as properties of linguistic entities. But there are, as we have stressed, other manifestations of believing that are not restricted to language users. One consequence of the false-kibbles experiment might be that, in a subsequent trial, the dog doesn't just head for the kibbles, real or facsimile. He may approach them with suspicion, sniff around them, turn them over. He behaves in every way like someone who does not yet believe that what is in the bowl is edible kibbles, or like someone who mistakenly believed on the previous go-round that what

16. ''Thought and Talk,'' in *Mind and Language*, ed. Samuel Guttenplan (New York: Oxford University Press, 1975), pp. 22–23.

155

was in the bowl was edible kibbles and has "learned" from that mistake.

If we allow that speech acts are not the only or overriding evidence for believing and that believing is not peculiar to language users,[17] then we can see more clearly what precisely is the claim being made by the disquotation principles, modified or unmodified. Both principles claim that certain speech acts on given occasions, speech acts of sincere, reflective, non-conceptually-confused language users, are *very privileged* behavioral markers of belief—that there is an essential connection between the assents of such agents and believing. Indeed, the disquotation principle seems to say, or rather imply, that, where such assents are available, one need not look at further behavior to determine what an agent believes. But, given the possibility of entertaining the decoupling of speech acts from believing, the further question might be raised whether, even among language users, however sincere, the speech act of assenting is so privileged a behavioral indicator of belief. Psychological theories tell us that our avowals about our own states, our desires and fears, about the objects of our affections and disaffections, are often unreliable; that is, the sentences we assent to, however sincerely, that purport to describe such states and affections are often poor and perhaps contrary indicators of what we believe. These are not cases of deliberate deception, of insincerity, as ordinarily conceived; they do have to do with those deeper notions of self-deception and false consciousness that I need not go into here. Such claims cannot be made if one supposes that there is an essential connection between sincere assenting and believing, which would make other indicators of belief evidentially superfluous, given sincere assent. But not all instances of incoherent behavior need be explained by self-deception or the like.

Consider an agent who sincerely assents to the true sentences of arithmetic and to their verbally given applications. He says, sincerely, "yes," to the whole range of sentences of arithmetic like '$2 + 2 = 4$', '2 oranges and 2 oranges make 4 oranges', etc. He never trips up in his assentings. But his nonverbal behavior is incoherent with his verbal behavior. He never makes correct change. If you ask him to give you a certain number of things, he never gives you that number. His non-

---

17. The disquotation principle (1) is a conditional that takes assenting into believing. Kripke, "A Puzzle about Belief," n. 10, also takes the converse of (1) to be reasonable: If a speaker of L believes that p and 'p' is a sentence of L, then he assents to 'p'. But surely we do not assent, perform a speech act, correlated with each of our beliefs. Even if we replace 'assents' by 'would assent' the converse is untenable unless we restrict the principle to an agent's own reports of his beliefs and assents.

verbal behavior also discloses that he is not in some other modulo. Are his sincere assents *sufficient* for ascribing beliefs to him? Are they overriding? Shouldn't his nonverbal behaviors count as indicators or counterindicators of belief? Here his nonverbal behavior is "cognitively dissonant" with his verbal belief claims.

Against this background we can more readily see the role of the disquotation principles. But first I will make an unargued claim about the nature of rationality, broadly conceived. It is a commonplace view, provided one can shift one's focus away from defining rationality narrowly in terms of the "propositional" analogues of consistency of sentences and closure under logical consequence—a focus that has dominated philosophical discussions of rational belief. We will say that an agent is broadly rational if *all* his behavioral indicators of belief are "coherent" with one another. For example, the agent's narrowly rational assentings will be coherent with his choices, his bets, and a whole range of additional behavioral indicators of belief. I will not presume, nor am I able, to provide a fleshed-out theory of rationality as "coherent" behavior, but the notion is familiar. We may say of someone who avows that he loves another, yet treats cruelly the one he claims to love and does not also avow that he is being cruel in order to be kind, that his behavior is incoherent, "dissonant"; that he is being "irrational." We say that he is "irrational" although he is not logically irrational in the narrow sense; he never says sincerely, "I love *A*" and then immediately says, "I don't love *A*." He would deny the latter if asked.

Agents who become aware of their own incoherent behavior may try to "rationalize" their behavior, make it coherent, get a better fit. On such awareness, an agent like the ambivalent lover may no longer assent to 'I love *A*', or he may alter other kinds of behavior that are belief indicators. He may even argue that the concept of love is confused. That does not exhaust the possibilities.

Against this larger notion of rationality we may see both disquotation principles (1, 1′) as asserting a privileged and *overriding* status to certain speech acts as belief indicators in language-using agents. They discount, for language-using agents, the role of other indicators as evidence for belief. Both disquotation principles may even be seen as already presupposing "rational" belief in the larger sense on the part of agents—that sincere assenting will or must be coherent with other belief indicators. If assenting sincerely is the *sure* indicator of belief, then no other behavior can be evidence for an agent's possibly not being in the belief relation to the state of affairs described by the sentence to which he assents.

Against the background of the disengagement of assenting and believing, and the broader notion of rationality, it is possible to see what underlies my intuition: that I would say, if I were Pierre, that I was mistaken in claiming on a given occasion that I believed both that Londres is pretty and that London is not pretty despite my assent to the conjunctive sentential clause. Also, against that background we can see that the modified disquotation principle is informed by that explanation.

I will, in describing Pierre's predicament, use the modified disquotation principle and account for that use. Recall that Pierre, before departing for London, has reasonable empirical grounds for assenting to sentence (3) 'Londres est jolie'. It is, after all, possible that London is pretty, and so the assent goes over into a belief that London is pretty. After emigration and acquiring English he assents on reasonable grounds to (4) 'London is not pretty', which also goes over into a belief. Since proper names can be carried over from language to language and given that Pierre is logically rational in the narrow sense of rationality, he would also assent to the contingent sentence

(5)   'London' and 'Londres' name different things.

which carries over into a belief. Also to

(6)   London is different from Londres.

which, on the modified disquotation principle, does *not* carry over into belief.

It is important to notice that Pierre does not claim to have (5) and (6) among his *initial* stock of beliefs. It is his *very logical* rationality, his acceptance of Leibniz's law,[18] and his assenting to all the logical consequences of what he assents to that *lead* him to assent to (5) and (6) and therefore to claim to believe (6), an impossibility.

Pierre would of course not even *assent* to the sentences:

(7.1)   Londres is not pretty.
(7.2)   London is pretty.
(7.3)   Londres is not pretty and London is pretty.

Nor, given logical rationality, would he assent to the formal contradictions:

(8.1)   London is pretty and London is not pretty.
(8.2)   Londres est jolie et Londres n'est pas jolie.

18. By "Leibniz's law" we mean the principle that requires that, for identity to hold between an object *a* and an object *b*, *a* and *b* must have the same properties.

Those who do not clearly distinguish between *assenting,* which is a relation between a language user and a linguistic utterance on a given occasion, in a particular context, and *believing,* which is a more enduring relation between an epistemological agent and a state or states of affairs not necessarily actual, will fail to see the distinction between assenting on a given occasion to the utterance 'London is not Londres' and believing that London is not Londres. Those who keep the distinction clear may, I think, share my Berkeleyan intuitions, for we can see how a non-conceptually-confused wholly rational agent can come to *assent* on a given occasion to a sentence that describes an impossibility. And we can see how Pierre, on discovering that some contingent sentences to which he has assented are false—sentences like 'Londres is pretty' and " 'London' and 'Londres' name different things"—would be led to claim that his apparent belief in an impossibility was just that: apparent. This is not to deny some of the behavioral similarities of the purported belief relation to normal cases, even in the failed case. As Berkeley suggests, rational agents may in some respects act *as if* they had such beliefs, but they are only imagining that they have them. The supposition of modified disquotation is that, for the belief relation to *hold,* it must have a proper object, a possible state of affairs, for how otherwise is one to distinguish Pierre's other behavior, such as his "choice" behavior, from that of a wholly irrational agent? A wholly irrational agent is not merely *logically* irrational in the narrow sense in that, for example, he might assent to the sentence 'London is pretty and London is not pretty' where common reference is supposed for both occurrences of 'London'; he is an agent for whom linguistic and nonlinguistic belief indicators are persistently incoherent. There is a question here whether, with widely incoherent belief indicators, he has beliefs at all, for what work could "belief" be employed to do in such a case?

Pierre, recall, is logically rational in the narrow sense. Let us even suppose that he is also rational in the wide sense. He has been through years of therapy, or whatever it is that is required to induce coherence of belief behavior. Yet some of his behavior, and some of the intended outcomes of his purported beliefs about London, would be just like those of an irrational agent. If, for example, Pierre, in accordance with his assents, were to bet on a given occasion on Londres's being pretty and London's being not pretty, the *outcome* of his bet would be as self-vitiating as that of the wholly irrational agent. Yet purely *logical* considerations and the empirical justification available to him made that bet wholly "reasonable." Or suppose, given his assents, he "chooses" to move from ugly London to pretty Londres. He may behave in some respects as if there were such a choice, and as a rational agent he can *explain* his strange behavior. Given his assent to 'Londres est jolie' and to 'London is not pretty' and his bilingualism, it is logical for him to assent to the conjunction of those sentences. But the assent does not carry over into a belief.

Loosely speaking, we could describe this outcome as follows. Even where assent factors out of a conjunction, belief does not. This is not to suggest that, on any given occasion of assenting, assent always factors out of a conjunction. Witness the lottery paradox in which there is a winning ticket among some large number of tickets, say a million, numbered accordingly. An agent assents to each of '1 won't win', '2 won't win', and so on, up to a million, as well as assenting to 'A million tickets is all there are' and assenting to 'There is one winning ticket'. If, as we are assuming, he is wholly logically rational, he will not even *assent* unqualifiedly to the conjunction of those sentences; so the disquotation principle does not apply.[19] In the case of Pierre it does seem logically rational for him to *assent* to the conjunction 'London is not pretty and Londres is pretty' on the occasions when he assents to each of the conjuncts, but, in accordance with modified disquotation, the assent to the conjunction does not carry over into belief.

It should now be clearer what some of the considerations are that explain the intuition I and others have that, if I were Pierre, I would say that I was mistaken in claiming that I believed Londres was not the same as London, and, similarly, my intuition that I would disclaim having believed that Hesperus was not Phosphorus when I learned that they were the same. For beliefs, whatever they are, are supposed to figure in our intentions: to guide choices, to guide action in the large, not just speech acts. They are supposed to guide us in achieving certain ends. But if a sentence that I assent to describes an impossible state of affairs, whether or not I know or believe that it is such a description,

19. See Henry Kyburg, "Conjunctivitis" in *Induction, Acceptance, and Rational Belief*, ed. Marshall Swain (Boston: Reidel, 1970), pp. 55–82. There, given the lottery "paradox," Kyburg rejects the principle that 'believes' or 'accepts' factors out of a conjunction, but, as we have noted, neither would the assents; and the lottery case is therefore not a counterexample to the disquotation principle. Of course there is the question whether an agent would even *assent* to each of the conjuncts. 'Assent' as used here is a yes-no affair. 'Accept' may be weaker, suggesting degrees. See my paper "A Proposed Solution to a Puzzle about Belief."

The lottery "paradox" shares features with the puzzle generated by an agent's reasonable "belief," given nonomniscience, that, of his large number of "beliefs," one is false. Such an agent might assent to each of 'p_1', 'p_2', . . . , 'p_n' and also to '$\exists x(x$ is false)', where the variable ranges over these very sentences, but he would not assent to their conjunction if he is logically rational.

there is no way I will be served in achieving any ends. Such a sentence is not even a candidate for truth. Indeed, my total behavior in that context will be rendered incoherent, like that of a wholly irrational agent in the wider sense of rationality. I can, however, *explain* how my very rationality accounted for my various assents, but my final assent to a sentence false in all possible circumstances will not carry over into a belief.

11. *Spinoza and the Ontological Proof*

This paper appeared in A. Donagan, A. N. Perovich, Jr., and M. V. Wedin, eds., *Human Nature and Natural Knowledge* (Dordrecht: Reidel, 1986), pp. 153–166. That volume honored Marjorie Grene, who has long deserved our admiration. It overlaps with an earlier essay, "Bar-On on Spinoza's Ontological Proof," presented at a conference held in Jerusalem in 1977 marking the three-hundredth anniversary of Spinoza's death. The latter appeared in *Spinoza: His Thought and Work,* ed. N. Rotenstrelch and N. Schneider (Jerusalem: Israel Academy of Sciences and Humanities, 1983), pp. 110–120. ■

I should like in this paper to characterize the "ontological proof" for the existence of God. In so doing I will discuss Harry A. Wolfson's[1] analysis of the proof and argue that his characterization is not, as has been claimed, wholly mistaken, but is insufficient. I will also argue that the more adequate characterization here proposed gives us a good account of Anselm's and Descartes's proof, but that, on that account, there is no "ontological proof" in Spinoza or at least not in the text where it has usually been thought to be located. In conclusion, I will touch upon the confusion of metaphysical notions and epistemological notions, such as that between conceivability and possibility, which has conflated quite different sorts of arguments under the heading "ontological proof." It is clear from these preliminary remarks that the term 'proof' is being used in the wide sense of argument and not in the narrow sense of valid or successful argument. Nor will I be concerned with the validity of those arguments that qualify as ontological.

I

In *The Philosophy of Spinoza,* Wolfson,[2] after a lengthy discussion of texts, concludes:

> **W1** . . . none of the ontological proofs in their various forms as given by its three main exponents, Anselm, Descartes, and Spinoza, prove directly that God exists. What they prove is that the existence of God *is known to us by a certain kind of immediate knowledge.* Their various proofs can be reduced to the following syllogism:
>
>> If we have an idea of God as the greatest, or as the most perfect, or as a self-caused being, then God is immediately perceived by us to exist.
>>
>> But we have an idea of God as the greatest, or as the most perfect, or as a self-caused being.
>>
>> Therefore God is immediately perceived by us to exist.

It should be noted that in the quoted passage and throughout, Wolfson uses 'is perceived by us' and 'is known to us' interchangeably. What Wolfson says in the quoted passage is that the ontological proof is not a direct proof of God's existence. What he can be understood to

---

1. *The Philosophy of Spinoza,* 2 vols. (New York: Meridian Books, 1958).
2. Ibid., p. 176. Italics added.

be saying is that it is an indirect proof, but indirect proofs cannot be excluded as not genuine. *Reductio ad absurdum* is often called "indirect proof." However, I am not suggesting that it is this logical mode of indirection that is the basis of Wolfson's claim. It is rather, for want of some standard locution, what I will call an epistemological mode of indirection. In the quoted passage, Wolfson tells us that what is proved *directly* is that the existence of God *is known* to us by a certain kind of immediate knowledge. Suppose Wolfson's claim is correct; that what is proved directly is, in part, the proposition

It is known to us that God exists.

Now a plausible analysis of statements like

It is known that *P*.

or

x knows that *P*.

is that a necessary condition for their being true is that *P* is true. It is incoherent for *P* to be known and *P* to be false. There are some attempted analyses that challenge the claim that the truth of 'It is known that *P*' entails the truth of *P*. Nevertheless the more generally accepted analysis is persuasive, and it does take

P is known and *P* is false.

as inconsistent.

If therefore we can prove *directly* that

It is known that God exists.

as Wolfson is claiming in the quoted passage, then it requires one further step to arrive at

God exists.

The validity of that further step is not peculiar to the theological content.

Now, we do not usually demonstrate propositions by first demonstrating that they are known and then showing that, as a consequence of their being known, they are true. But it is interesting to examine why, for it illuminates Wolfson's claims. What would a demonstration of, for example,

It is known that the sum of the angles of a triangle consists of two right angles.

consist in? How could that statement be established? Well, one obvious way is for some epistemological subject to produce a proof. On the common analysis of 'It is known that *P*' it is not merely a necessary condition that *P* be true, but the knower, the epistemological subject, must be able to justify his claim. In mathematics that justification usually and perhaps always consists in providing a proof. I say "usually" because some mathematical realists sometimes claim that they have mathematical knowledge intuitively and that such intuitions may count as justifications. The claim is abetted by some of Gödel's results, for one can interpret those results as having the consequence that there are mathematical truths for which there never will be a proper proof, which are undecidable, but for which intuitive knowledge might be claimed. But, setting those considerations aside for the moment, let us suppose that, for mathematical propositions, demonstrating that *P* is known consists in showing directly that someone has in fact produced a proof of *P*. But for that, we would, in effect, have to display the proof. And displaying a proof is simultaneously a proof of *P* and a proof of

It is known that *P*.

If our interest is in proving *P*, then the further step of asserting that I know it is otiose.

Now, what counts as a mathematical proof can be made explicit by setting out canons or rules. There are, of course, in contemporary foundations of mathematics, disagreements about appropriate canons, constructivist versus nonconstructivist canons, for example. Still, whatever one's position on rules of proof, it is agreed that they must be wholly explicit, so that a proof can be checked against them. There can, therefore, be a mechanical check against the canons, and there is a sense in which we can say of a mathematical demonstration that it is possible for some *P* that it be not merely provable but proved, yet not known. A machine can be programmed to prove simple and even moderately complex theorems, some of which may be genuinely in doubt. If no epistemological subject examines those proofs and printouts are discarded, *P* remains proved but not known. I deny (a position that I will not defend here) that any machine capable of generating such proofs can also be said to know that *P*.

Mathematical demonstrations, like most demonstrations, are not routed through an epistemological step of showing first that they are known. The cosmological arguments, the arguments from first cause to the existence of God, are such that validity is claimed for them independent of any step about what is conceived or known. What distinguishes the ontological argument is the indispensability of the epistemological subject in arriving finally at the conclusion that God exists. In this respect, the ontological argument is like the *cogito*. The conceiving, understanding, knowing subject is indispensable to the demonstration. In the ontological argument, what is concluded directly is that something is known to us, i.e., the existence of God. As noted above, it is a trivial step from

It is known that God exists.

to

God exists.

What makes the ontological proof indirect is that it is routed through epistemological claims about what we conceive, understand, or know.

Although Wolfson is not uniformly clear about what was or was not intended in the ontological argument, it is quite consistent with his view that the ultimate intention of at least some of the philosophers cited was to prove that God exists and to present an argument in which the statement of God's existence is the conclusion in the strictest sense of the term.

Indeed, Wolfson says as much in a passage immediately preceding the one already quoted **(W1).** In that passage "God exists" is finally in the conclusion. Wolfson[3] says,

W2 Truly speaking, if the ontological proof were to be put into a syllogistic formula in such a way as to bring out its *entire* force, it would have to be as follows:

Everything which is immediately perceived to exist exists.

God is immediately perceived to exist.

Therefore, God exists.

The second premise, "God is immediately perceived to exist," is the conclusion of Wolfson's first-quoted syllogism **(W1).** In **W2** we have, in accordance with our gloss, the move from its being known or perceived that God exists to the conclusion that he does. The two syllogisms taken together form a sorites in which "God exists" *is* finally a conclusion in the strictest sense of the term. But it is the first-quoted syllogism **(W1),** the movement from having an idea or concept of God to knowing that the *ideatum* to which it is adequate exists, that marks the peculiarity of the ontological argument.

3. Ibid. Italics added.

167

Having defended Wolfson so far, I should like to go on to discuss a deficiency in his characterization, for I am in agreement that the account is not wholly adequate. It is, rather, insufficient. Unless Wolfson tells us more explicitly what the special features of the "certain kind of immediate knowledge" are, too many arguments will count as "ontological." Now, the syllogism in **W1** that Wolfson says displays the structure of the ontological argument is a hypothetical syllogism of which the conditional premise is

C   If we have an idea of God as the greatest, or as the most perfect, or as a self-caused being, then God is immediately perceived by us to exist.

The antecedent of C is generally taken as noncontroversial. It just says that we have the concept, i.e.,

We have an idea of God as the greatest, or as the most perfect, or as a self-caused being.

from which it follows by *modus ponens* that

God is immediately perceived by us (or known by us) to exist.

The deficiency in Wolfson's characterization is of course in failing to say more about the conditional premise that must be true if the conclusion is to follow. He does say in **W1** that the immediate perception, the immediate knowledge, is of "a certain kind," but he seems not to distinguish that kind from claims to immediate knowledge, such as that of the mystic or other psychological claims of immediacy. Wolfson's failure to differentiate kinds of immediate knowledge is seen from his remarks on Anselm's[4] response to Gaunilo's objections. Anselm's response appeals to Gaunilo's faith and conscience, and Wolfson[5] says:

Is it not possible that in appealing to faith and conscience Anselm is really invoking the argument from revelation as attested by the tradition by which the existence of God is established as a fact of immediate personal experience? Such an argument is common in Jewish philosophy, and it may be considered as partly psychological, in so far as the proof from revelation derives its validity from the fact that it is an immediate experience, and partly his-

---

4. *St. Anselm's Proslogion with a Reply on Behalf of the Fool by Gaunilo and the Author's Reply to Gaunilo*, ed. M. Charlesworth (Oxford: Clarendon Press, 1965), p. 169.
5. Ibid., p. 171.

torical and social, in so far as the truth of the fact of revelation is attested by an unbroken chain of tradition universally accredited within a certain group.

It is correct to point out that the texts of Anselm and Descartes do not support Wolfson's suggestion that the kind of immediate knowledge referred to in the schema of the ontological argument **(W1)** is akin to revelation. But that does not wholly vitiate Wolfson's account. Rather, it shows it to be insufficient.

The insufficiency of Wolfson's characterization is then in his failing to say more about the conditional premise. He allows that the premise might be established by revelation or by undifferentiated claims of psychological immediacy. In any case, an argument of the form of Wolfson's syllogism, in which the conditional premise is unsupported by further rhetoric of a logical or rational sort, should not be included among the ontological arguments. The assertion of the mystic who insists without further explanation that to have the vision of God is to know that he exists *simpliciter* is not in the spirit of those arguments of Anselm and Descartes that have been singled out as ontological. In stopping where he does, Wolfson does seem to classify among the ontological proofs those in which further rational support or justification need not be given for the conditional premise. But what is crucial in Descartes's and Anselm's arguments and what marks them as ontological is that they *do* attempt to give rational support to the conditional premise. They try to persuade us that a rational intelligence cannot deny that premise without incoherence. A rational intelligence cannot have the concept of the greatest or most perfect being and yet fail to know that God exists, for such a conjunction, they claim, leads, if not to outright contradiction, then perhaps to what G. E. Moore has called a pragmatic contradiction. A *fool,* of course, devoid of some rational powers, may fail to appreciate the incoherence and, therefore, in the absence of faith and revelation, may say sincerely in his heart that there is no God.

If Wolfson is suggesting, and he seems to be, that the conditional premise that goes from the idea of God to immediate knowledge of God's existence is the same kind of certain knowledge that the mystic claims, then, as has been pointed out, there would be no accounting for some of the texts that purport to give an ontological argument, for those texts *do* attempt to give support for the conditional premise of a rational kind. When Anselm appeals to Gaunilo's faith and conscience, it is surely an appeal to the faith and conscience of a man guided by reason. Gaunilo is no fool.

My disagreement with Wolfson's critics is, therefore, on where to locate Wolfson's failure. It is the rational support for the conditional premise that Wolfson has failed to see as the special feature of an ontological argument. That rational support is manifestly clear in Descartes.[6] Consider the passages from the Fifth Meditation.

> **D1**   It is certain that I no less find the idea of God, that is to say, the idea of a supremely perfect Being, in me, than that of any figure or number whatever it is, and I do not know any less clearly and distinctly that an actual and eternal existence pertains to this nature than I know all that which I am able to demonstrate of some figure or number truly pertains to the nature of this figure or number. . . .

and

> **D2**   . . . when I think of it with more attention I clearly see that existence can no more be separated from the essence of God than can its having its three angles equal to two right angles be separated from the essence of the rectilinear triangle, or the idea of a mountain from the idea of a valley; and so there is not any less repugnance to our conceiving a God that is a Being supremely perfect to whom existence is lacking (that is to say, a certain perfection is lacking) than to conceive of a mountain which has no valley.

In the first quotation **(D1),** Descartes simply says that he finds the idea of God in himself and he knows from the idea clearly and distinctly that an actual and eternal existence pertains to this nature *and that he knows that no less clearly and distinctly than he knows that some property he is able to demonstrate about a figure or a number pertains to the nature of that figure or number.* Now Descartes is claiming an immediacy for the connection that those who criticize Wolfson for placing a premium on immediacy have failed to appreciate. There is in the quoted passage not even a *reductio* argument. What Descartes is saying is that, without further demonstration, and from the idea of God, he knows that God exists with *at least* the clarity and distinctness with which he knows a demonstrated mathematical truth. He is claiming that the knowledge, albeit immediate, is rational.

Again, in the second quotation from Descartes, although there is the hint of a *reductio,* the immediacy of the move from a rational

6. *Philosophical Works*, trans. E. Haldane and G. Ross, 2 vols. (New York: Dover, 1955), pp. 180–181.

subject's idea to knowledge of existence is highlighted. We cannot, says Descartes, have the idea of a mountain without at the same time having the idea of a valley. Similarly, he says, it is repugnant (which I read as "repugnant to reason") to conceive of a God to whom existence is lacking. A reasonable man cannot have the concept of God and yet fail to know immediately that God exists, and that is, after all, Wolfson's conditional premise.

What I have claimed in the foregoing is that it is correct to fault Wolfson for being untrue to the texts, but not because Wolfson claims that the conclusion of the ontological argument is that God is known to us by a certain kind of *immediate knowledge*. It is rather because of Wolfson's failure to say something more about the conditional premise, his failure to distinguish the "immediacy" of the knowledge claimed in the ontological argument from that claimed for revelation or for other kinds of psychological immediacy. The "certain kind of immediate knowledge" by which a knowledge of God is claimed in the ontological arguments should have been singled out as *rational* intuition. Having the idea, grasping the essence of God, is supposed to lead us to know, with clarity and distinctness, that he exists. Given that the conditional premise asserts immediacy between antecedent and consequent, we cannot press the sorites back further in an obvious way. There is no further premise. That is, after all, what immediacy entails. But we can bolster support where rational intuition is somewhat blunted. We can get such a man to see it for himself in a manner akin to Socrates' efforts in the *Meno*. We can point to an analogy with mathematical insights or with concepts that entail each other, such as the concepts of mountain and valley. There are *reductio* supports; we can try to show that denying the conditional leads to incoherence.

According to Anselm,[7] "No one in fact, *understanding* what God is, can conceive that God does not exist." Since anyone who claimed he could not conceive *not-P* would also claim to know *P,* Anselm's assertion fits Wolfson's schema of the conditional premise in **W1.** More precisely, Anselm's claim entails Wolfson's conditional premise, for the knowledge claimed by Anselm is not merely that the nonexistence of God is known to be false; propositions can be known to be false yet their truth still be conceivable. What he is asserting is that an epistemological subject, upon understanding what God is, will know that God exists with such certainty that conceiving of his nonexistence would be incoherent. Anyone who fails to make the connection is accused by Anselm of lacking "a rational mind" or of being

---

7. *St. Anselm's Proslogion,* chaps. III, IV.

"stupid and a fool." Here we see the departure from Wolfson. Stupidity, foolishness, or irrationality have never been claimed as impediments to mystical insight or revelation.

II

The question remains whether there is in Spinoza's writings an ontological proof as here characterized. Let us even suppose that Spinoza has succeeded in showing that God and substance are identical. Is there an argument that goes from the idea of God, or substance, and the claim that from that idea it is known that God or substance exists, ultimately to the existence of substance? Is there supporting rhetoric for the conditional premise, such as mathematical analogies, reductions, and the like?

Now, it is possible that from Spinoza's complete texts one might be able to construct such an argument. But there is one segment of the *Ethics*[8] that has been singled out by Wolfson and others as constituting an ontological argument. Broadly speaking, it is all of Part 1 terminating in Proposition XI, which says that God or substance exists necessarily. My claim is that in Part 1 of the *Ethics,* or indeed elsewhere in the *Ethics,* there is no ontological proof to be found as here characterized.

The *Ethics* is patterned after Euclid's geometry. At the outset that geometry presupposes the existence of certain objects such as points and lines. The definitions and axioms define some of their essential properties and some of the relations between those geometrical objects. Analogously, Spinoza in the *Ethics* begins with the presumption of the existence of things, substance, modes, God, and some relations between them, including the relation of identity. The *Ethics* is largely an unpacking and elaboration of those relationships. Spinoza's definitions with which he begins are like Euclid's "real" definitions. It is not merely "ideas" that are being defined, which must *then* be shown to correspond to their ideata. Euclid's definitions were understood as telling us what objects geometry is about, and they define or mark out some of their essential properties. Spinoza's eight definitions are about things, finite and infinite, substance, modes, God, free things, compelled things, and so on. It is *supposed* simultaneously that we have ideas of those things that are true to them. That those ideas are adequate to their ideata is never in question. Some of those

8. B. Spinoza, *Selections,* ed. J. Wild (New York: Charles Scribner's, 1930).

presumptions are spelled out in the axioms. Axiom I already tells us that everything is either in itself or in another. Given the definitions, existence of examples of each kind are presupposed. For example, Definition VI supposes that there are things that exist from the necessity of their own nature. They are free things. Compelled things are those that are determined to existence by another. Note that the demonstration of Proposition VII, the assertion that it pertains to the nature of substance to exist, is not a *proof* of the existence of substance. That is already supposed. It purports to show that the existence of substance, unlike other existences, follows from its own nature. Indeed, Spinoza wonders whether a demonstration of Proposition VII is even required. He says in the second Scholium of Proposition VIII,

> If men would attend to the *nature* of substance they could not entertain a single doubt of the truth of Proposition VII. Indeed, this proposition would be considered by all to be axiomatic and reckoned among common notions.

As for Proposition XI, which is usually singled out as the conclusion of the "ontological proof," i.e.,

> God or substance consisting of infinite attributes . . . , necessarily exists.

it is a direct consequence, and instance, of Proposition VII. One cannot locate in Part 1 of the *Ethics* any demonstration that from the *idea* or *concept* of God it follows that he is known to exist or that he exists. What is asserted in Proposition XI is that he exists in a certain way—necessarily. His existence *simpliciter* is never in question.

The demonstrations that substance and God exist in a certain way, i.e., necessarily, may misleadingly suggest the form of an ontological argument to those who do not trace it from the definitions. The Scholium of Proposition VIII does say that "if men would attend to the nature of substance they could not entertain a single doubt of the truth of Proposition VII" i.e., that the existence of substance follows from its nature. Proposition XI does ask us to conceive, if we can, that God does not exist. But those urgings are used not to show that he exists or that he is known to exist but to show that we would be forced to deny that his essence involves his existence, which is one of the essential or defining properties of God or substance. To argue that the existence of substance or God follows from its nature is quite a different matter from arguing that the existence of substance or God follows from our concept of it. That our idea of God has an ideatum (as all ideas do for Spinoza) and is furthermore adequate to its ideatum is, at least in the *Ethics,* taken as true at the outset. In summary, although Wolfson was correct but incomplete in his characterization of an ontological argument, he was wrong in claiming that such an argument could be found in the *Ethics.* The place for an ontological argument in the *Ethics* would have been in support of some of the axioms and definitions.

III

I should like in conclusion to say something about the uses of modal logic in our understanding of proofs that have been classified as ontological. Some and perhaps much of the recent revival of interest in such proofs stems from the revived interest in modal logic and the belief that ontological arguments can be translated into modal arguments. I do not propose to go into all of those various efforts, although the extent to which those efforts succeed, fail, or are misapplied is itself illuminating. I should like just to mention two such attempts, those of Charles Jarett and of Alvin Plantinga.

In a paper, "Spinoza's Ontological Argument," Jarett[9] argues that if we interpret the first part of the *Ethics* as claiming that

(1)   It is possible that God exists.
(2)   Necessarily (If God exists then he exists necessarily)

it follows in one of the strong modal systems (with some constraints on what is known as the Barcan formula) that

(3)   Necessarily God exists.

I do not propose to discuss whether such an argument as this can be culled from the *Ethics.* Like many such attempts, as it stands, it is not an ontological argument as we have characterized that genre. It does not have the requisite epistemological features. There is no conditional that takes us from what is *conceived* to what is known or to what exists. There is no argument to warrant the identification of 'conceivable' with 'possible'. To convert the ontological argument as here characterized into a modal argument like the one presented above supposes that certain epistemological notions can be translated into logical or metaphysical ones; that, for example, to say of something

9. *Canadian Journal of Philosophy,* V (1976): 685–691.

that it is conceivable is to say that it is possible, and to say of something that it is inconceivable that it not be the case is to say that it is necessary.

Alvin Plantinga,[10] on the other hand, in his modal transformation of Anselm's argument, affirms the legitimacy of such a transformation. He asserts, without support from Anselm's texts, "And when [Anselm] says that a certain state of affairs is conceivable he means to say (or so at any rate I shall take him) that it is a logically possible state of affairs—possible in our broadly logical sense. So for example . . . . 'God's existence in reality is conceivable' is more clearly put as 'It is possible that God exists in reality'.''

On the most common and plausible use of such epistemological concepts as "conceivable," there is no justification for the identification. Conceivability requires an epistemological subject, someone to do the conceiving. A proposition may be taken as conceivable to an epistemological subject if it is consistent with the subject's beliefs or, more weakly, if he believes it to be consistent with his beliefs. On such an account there are powerful counterexamples to the identification of epistemological concepts with logical or what Kripke has lately called "metaphysical" modalities. A mathematical claim, for example, if true, is necessarily true, and, if false, is necessarily false. It is not controversial to suppose that, when a mathematician conjectures that P where P is a mathematical proposition, he finds it conceivable that P. Yet P may be false and hence impossible. Indeed, such was the case with at least one of Hilbert's list of conjectures. Or consider the more mundane case where someone finds it conceivable that Hesperus is not the same as Phosphorus, yet, as an empirical matter, "they" are found to be identical. But if x and y are identical, "they" are necessarily identical; hence it is impossible that they not be the same, even though it was conceivable to some epistemological subject, in ignorance of the discovery of their identity, that Hesperus was different from Phosphorus.

Furthermore, and apart from such counterexamples, modal arguments like Jarett's and Plantinga's, if valid, would remain valid in the absence of *any* epistemological subject or any idea (or concept) of God altogether. Modal arguments for the existence of God, arguments that purport to prove God's existence from his essence, are interesting in their own right but are not to be conflated with paradigm examples of the ontological argument such as Anselm's and Des-

10. *The Nature of Necessity* (Oxford: Clarendon Press, 1974), p. 199.

cartes's. Perhaps "epistemological proof for the existence of God" would have been a more appropriate designation for Anselm's proof. What remains to be explored is the set of presuppositions that connect those metaphysical and epistemological concepts that have led to the confusion and conflation of the two kinds of arguments.

12. On Some Post-1920s Views of Russell on Particularity, Identity, and Individuation

In June of 1984 the Fondation Singer-Polignac sponsored a colloquium on the merits and limits of methods of logic in philosophy. Maurice Boudot in "L'Individuation, Vrai ou Faux Problème?" and I were concerned with the evolution of Bertrand Russell's views on particulars, universals, individuation, and indiscernibility. The paper here printed is an edited version of my contribution.

It was a marvelous conference organized by Professor Jules Vuillemin of the Collège de France, who also edited the proceedings, *Mérites et Limites des Méthodes Logiques en Philosophie* (Paris: J. Vrin, 1986). ∎

Bertrand Russell's writings in metaphysics and epistemology after 1920 suffered from nonbenign neglect, and not for the reason that those efforts were mere elaborations of earlier views. Quite the contrary. Russell was one of those uncommon philosophers who readily acknowledged that susceptibility to error was an ongoing feature of any inquiry and that philosophical theorizing was not immune. He frequently changed his mind without apology but not without reasons. The preeminent cause of neglect of the later work was a widespread belief that it was of less interest and of little consequence. But that is unjustified.

Maurice Boudot[1] has traced a radical shift in Russell's views on the nature of particulars and universals, individuation, and indiscernibility. He has raised questions about the viability of Russell's revised ontology and touched on its consequences for logic. I would like to expand on Professor Boudot's discussion and in so doing to indicate (1) what remains constant in the evolution of Russell's ontological views against which his later theories seem to be natural outcomes; (2) some criticisms of Russell's earlier views, not mentioned by Boudot, which may be seen as catalyst for the change; (3) some consequences the revised theory has for Russell's logic and its interpretation; (4) comments on questions raised by Boudot about Russell's later views, where Russell appears even more clearly than formerly as a harbinger of contemporary theories of demonstratives and ordinary proper names. In so doing I will avail myself of Boudot's many interesting insights.

I

There is in Russell's ontological development a certain constancy of plan that is preserved in the evolution of his theories. It can be summed up with some quotations from "The Philosophy of Logical Atomism."[2]

> One purpose that has run through all that I have said, has been the justification of analysis, i.e., the justification of logical atomism, of the view that you can get down in theory, if not in practice to ultimate simples out of which the world is built, and

---

1. "L'Individuation, Vrai ou Faux Problème?'', in *Mérites et Limites des Méthodes Logiques en Philosophie,* ed. Jules Vuillemin (Paris: J. Vrin, 1986), pp. 49–67.
2. In *Logic and Knowledge,* ed. Robert Marsh (New York: Macmillan, 1956), pp. 270–271. The original was published in the *Monist* (1918).

that those simples have a kind of reality not belonging to anything else. Simples . . . are of an infinite number of sorts. There are particulars and qualities and relations of various orders . . . but all of them . . . have in their various ways some kind of reality that does not belong to anything else. The only other sort of object you come across is . . . facts, and facts . . . are not properly entities at all in the same sense in which their constituents are. . . .

Another purpose . . . is the purpose embodied in the maxim called Occam's Razor. . . . take some science, say physics. You have there a given body of doctrine, a set of propositions expressed in symbols . . . but you do not know what is the actual meaning of the symbols that you are using. The meaning they have *in use* would have to be explained in some pragmatic way: but their *logical* significance . . . is a thing to be sought, and you go through . . . these propositions with a view to finding out what is the . . . smallest empirical apparatus—or the smallest apparatus not necessarily wholly empirical—out of which you can build up these propositions. . . . that problem . . . is by no means a simple one. . . . It is one which requires a very great amount of logical technique. . . .

The shift in ontology from "The Philosophy of Logical Atomism" to *Human Knowledge: Its Scope and Limits*[3] may be seen as an evolution consistent with the earlier statement of purpose. Atomism remains. Simples are still being pursued, but the "apparatus" has been reduced. There now is only one sort of simple: qualities, i.e., qualities or "elementary" properties. Particulars have been redefined, but in a way that advances Russell's statement of purpose. Let us review the change.

A feature of the earlier views that troubled Russell was the nature of his particulars. Particulars are simples of a sort different from qualities. They are the bearers of qualities, but little more can be said about them. Russell acknowledges that his particulars are much like the individual substances of traditional metaphysics in that they are subsistent unities; however, unlike the traditional individual substances, they are very short-lived and do not persist. They are, furthermore, individuals that can be properly named but not described. They enter into direct acquaintance. Particulars can be known, independent of any proposition of which they might be constituents. Indeed, propositions

---

3. New York: Simon and Schuster, 1948.

and propositional functions are viewed as parasitic upon particulars. Particulars compose the pool of candidates for converting atomic functions into propositions, which is the first step in the constructive enterprise.

Given the transient character of particulars, they are "named" on the occasion of their occurrence by a demonstrative and an act of ostension of the perceiver (or epistemological subject), and to that extent those "names" are not syntactical elements of a *public language* with fixed value. 'This is red' is an atomic sentence where 'this' on a given occasion refers to an elusive transient subject.

In the later work a substantial part of the apparatus remains intact. But particulars are no longer among the simples. They will retain some of their earlier features, such as unity. To quote from the *Inquiry,* Russell says that 'this is red' is not a subject-predicate proposition, but is of the form 'redness is here'; . . . 'red' is a name, not a predicate; and . . . what would commonly be called a 'thing' [a particular] is nothing but a bundle of coexisting qualities such as redness, hardness, etc."[4] The troublesome empirical *inaccessibility* of his earlier (bare) particulars has been analyzed away. An advance in empiricism as well as Occamism.

Although for the earlier Russell qualities are also simples, his claim was that *understanding* a predicate that designates a property or quality requires "a different kind of act of mind" than that required for understanding a name. A *proper* name is "understood" in the act of naming, in its immediate reference to an object of acquaintance. Predicates in early Russell are not directly referential.[5] To understand a predicate-expression requires, as early Russell says, "bringing in the form of the proposition." Predicate expressions have the form of open sentences such as 'x is wise', and generate or "induce" either the set of wise things or the set of properties true of being wise. *No entity* is simply named by 'wise' (shades of Frege and unsaturatedness).

Atomic qualities remain among the later Russell's simples, but here, it seems, they differ from particulars not because "a different act of mind is required" for their understanding. We are acquainted with qualities directly. They are not the particulars of his earlier views because the latter must be transient, nonrecurring, and orderable spatially and temporally. Qualities may (but need not) appear in several locations at once and at different times in the same place. Space-time locations in the phenomenal field are also qualities, but they do not

4. *An Inquiry into Meaning and Truth* (London: Allen and Unwin, 1980), p. 97. Originally published in 1940.
5. "The Philosophy of Logical Atomism," pp. 204–205.

have such repeatable features. Complex, *non*atomic expressions such as '*x* is red or *x* is round' do not name and so may require for their understanding a ''special act of mind.'' If open sentences are seen as somehow signifying properties in some extended sense, that only serves to distinguish properties broadly conceived from atomic qualities. ''Red'' taken by itself (if redness *is* an atomic unanalyzable quality) simply names the quality red, as we learn from the above quotation.

Despite the apparent reversal of views, what we also have is a further advance consistent with Russell's avowed purpose. First in the empiricist direction. A quality is no longer an abstract object ''manifested'' in its perceptible instances. It *occurs* in each of those instances. The quality—the universal—presents no Platonic problem of *relating* its occurrent perceptible presentations to some abstract universal. Second, in the application of Occam's razor, no higher-order universal of redness is required in which the instances of redness ''participate.''

Viewed, then, against the earlier statement of purpose, Russell's later views are natural outcomes. He is still atomistic, i.e., reductive to basic simples. But, consistent with Occam's razor, the kinds of simples have been reduced. There is only one kind: nonabstract universals. The orders of simples have also been reduced. Empiricism has also been advanced. *Bare* particulars as well as *abstract* nonempirical ''qualities'' inaccessible to perception have been analyzed away.

Boudot draws our attention to the shifts in ontology that seem like radical reversals, but against a wider background we could see them as a very natural evolution of the program. The historical question is, what precipitated the shift? I am of course not wholly familiar with the vast Russellian literature between 1912 and 1950. But in the two central books, *Inquiry* and *Human Knowledge,* little is said to trace the source of the change. As Boudot points out, one is startled to find in the *Inquiry* passages almost verbatim from the 1912 paper ''The Philosophy of Logical Atomism'' in which the irreducibility of universals to particulars (and vice versa) was claimed, except now they are being used as a statement of a position to be repudiated.

II. Ramsey's Critique of Russell

It is a plausible hypothesis that a paper of F. P. Ramsey, ''Universals,'' largely a critique of Russell, which appeared in *Mind* in 1925 and later in *Foundations of Mathematics and Other Essays* in 1931,[6] was the catalyst for the shift in Russell's views. Russell does not refer to that paper in the *Inquiry* or in *Human Knowledge,* but perhaps he acknowledges it elsewhere. The connection is virtually inescapable. Ramsey's paper is worthy of discussion in its own right, but I will touch on only some of the arguments advanced to undermine Russell's earlier dualistic ontology.

First Ramsey argues that Russell has been blinded by the primacy of the functional notation in *Principia,*[7] where predicate expressions are taken as incomplete sentences, as in '*x* is red' or '*x* is red or square'. This led Russell to the position that understanding of predicates and understanding of names of individual particulars required different acts of mind; understanding of predicates requires, he says, "bringing in the form of the proposition." Since *genuine* names are referential without meaningful syntactical parts, predicates, which are incomplete symbols, with syntactic structure, are not genuine names. As Ramsey points out, the functional notation is efficient and convenient when for example one is dealing with complex expressions like '*x* is red or *x* is square'. Furthermore, for purely mathematical purposes where, given extensionality, we are concerned only with the sets induced by Russellian open sentences, one need not be concerned with the semantical range of higher-order variables. For mathematical purposes Russell ignores the role of names of attributes, where the name, taken by itself, could have a value to be identified not with the set of things having that attribute but with the attribute itself. Surely redness can be perceived and named directly without bringing in the form of the proposition '*x* is red'. What Ramsey concludes is that Russell has failed to attend to his own admonition not to be bullied by grammar or syntax, to which might be added notational convenience contrived for special purposes. In this case, the culprit is not the surface grammar of natural language but the grammar and notation of *Principia* itself, in conspiracy with extensionality.

A second argument originally advanced by Russell in support of his dualism is the familiar one that particulars can only be subjects whereas universals can serve as subjects as well as predicates.

From the formal point of view Ramsey restructures the claim as follows. Consider the sentence 'Socrates is wise' and replace 'is wise' by a variable, as in 'Socrates Φs'. All the propositional values of that function seem to be *about* Socrates as subject. But if we delete the

6. London: Methuen, 1932; reprinted, New York: Humanities, 1950.
7. With A. N. Whitehead, *Principia Mathematica,* vol. I, 2d ed. (Cambridge: Cambridge University Press, 1925). See "Philosophy of Logical Atomism," especially §III.

names of individuals from sentences that contain 'wise' as in 'x is wise' or 'neither x nor y is wise' there will be two *kinds* of sets of propositions that are its values: those that are about being wise itself and those that attribute wisdom to a particular. Ramsey argues that this asymmetry is superficial and arbitrary. If we are to take 'x is red' as meaning something different from 'red characterizes x' one must already have presupposed some special and distinguishable asymmetric role of the copula in atomic sentences for which no justification is given. The universal-particular distinction requires no *ontological* dualism. What we take as a particular, if not arbitrary, is guided by other considerations. Ramsey reminds us of Whitehead's ontology, in which events are basic and material objects and persons are converted into properties of certain classes of events, a view that Ramsey also favors.

Ramsey's conclusions are incorporated into Russell's later views. Qualities and particulars are of the same ontological type (but in Russell *not* of the same complexity). Russell's new particulars—those complete complexes of compresence—are constituted by simple qualities. They are (Russell hopes) distinguishable, unrepeatable, and hence countable as well as orderable in space-time. Indeed, consistent with Ramsey's ontological preferences, they may also be seen as made up of orderable identifiable events. Physical objects and persons are taken as specified concatenations and series of such particulars. The type-asymmetry between particulars and universals has been abandoned.

III. Russell's Logic As Related to His Later Ontological Views

Many questions arise with respect to how one might interpret the logic of *Principia* consistently with Russell's later ontology. If one takes as individuals, complexes of compresent qualities, then qualities and individuals are of the same type. As between them what is the relation of predication? To say of an individual that it is red would now be to envision some whole-part relation where redness is part of the complex particular. How then is the entire type structure to be reconceived? Indeed, since material objects and persons are series of such elementary complexes, the previous gap between logically constructed entities and basic individuals is considerably diminished. Does it all lead to a Goodman-like theory? Those are questions that take us some distance from the present discussion and I will not pursue them. But I would like to touch on Boudot's remarks about the Russellian definition of identity in *Principia Mathematica*. Boudot did not mention that Russell was quite aware of limitations of his definition and of its possible connections with and important differences from Leibniz's principle.[8]

Russell in *Principia Mathematica* notes that if properties are so broadly conceived as to include among the properties of an individual *A*, the property of being identical to *A*, then no special principle relating indiscernibility to identity is required. If a principle *is* required, then some narrower set of properties must be the basis for giving indiscernibility content. Suppose, says Russell, we take this set as those predicates that are *elementary* functions of individuals. Although he acknowledges that such a limited set may still be of a kind far wider than Leibniz intended, Russell notes that *even here* the second order definition of identity—where individuals with the same elementary properties are identical—would not be warranted without the *additional* assumption of the axiom of reducibility. That axiom says that, for every set of objects (including the unit set), there is *always* some elementary predicate unique to those objects. If a predicate is unique to a single object *A*, then if *B* has that predicate, *B is A*. Only then is the definition of identity, which quantifies over elementary properties, i.e., predicates of individuals, justified.

In Russell's later ontology, candidates for properties relevant to individuation are narrowed even further than what might be counted as predicates on a plausible semantics for *Principia*. They *are* Leibniz's properties, i.e., qualities, and then Boudot is quite correct. The postulated unique property may not be a quality. One needs something more than even the axiom of reducibility to justify the claim that the same complete complex of qualities guarantees uniqueness of the object. In the context of Russell's later theory, identity of individuals cannot be *defined* by indiscernibility. As he acknowledges, it is rather an inductive conclusion about the unlikelihood of there being two such complete complexes of compresence. And given that physical objects and persons are now constructed out of structured series of such complete complexes, the likelihood of there being two such wholly congruent structures is reduced even further. Shades of Mill's principle of the inverse variation of intension and extension!

Russell thus acknowledges that individuation is not absolutely guaranteed by indiscernibility. In the absence of a Leibnizian faith in God's sufficient reason, the identity of indiscernibles must be viewed as a powerfully supported *inductive* claim.

8. See *Principia Mathematica*, pp. 56–57.

We noted that in *Principia* Russell acknowledges that in the absence of the axiom of reducibility—about which he already has some reservations—identity would have to be introduced as an undefined relation. My speculation is that his reservations about reducibility, together with the claim that indiscernibility is an a posteriori inductive ground of individuation in his later ontology, would have made what is now the current practice of taking identity as primitive the preferable choice.

IV. Individuation and Reference

Russell's later books, *An Inquiry into Meaning and Truth* and especially *Human Knowledge: Its Scope and Limits,* are striking in their closeness to contemporary theories of direct reference, demonstratives, and proper names. In *Human Knowledge*[9] Russell asks,

> How shall we define diversity which makes us count objects as two in a census? We may put the problem in words that look different; e.g., "What is meant by a 'particular'?" or "What sort of objects can have proper names?"

Russell then sets out general conditions on particulars. A particular is a unity, nonrecurrent, spatially or temporally. It is, he hopes, also individuatable. His basic phenomenal particulars have all those features (or, as it turns out, it is very likely that they do). More specifically, they are complexes of qualities related temporally by compresence. Some of the qualities can appear in several places in the phenomenal field, but, given the absoluteness of space in the phenomenal field, simultaneous occurrences of qualities like red can be spatially distinguished. The complexes are complete. When composed of mental constituents, they are "total momentary experience." But such complexes need not require a nervous system. Recorders and cameras will do.

Having set out the nature of these basic particulars, Russell says, "With these preliminaries let us examine the question of proper names." Then instead of bringing in at this juncture the familiar demonstratives, the logically proper names of his earlier views, he goes on to give ordinary proper names a central role that denies that they are always as a matter of practice reducible to definite descriptions.

9. Chap. VIII of pt. IV, "The Principle of Individuation." The quotation is from page 292.

There are no hedges here, as in earlier Russell, about using the example of such an ordinary name as 'Caesar' with the caveat that, in fact, 'Caesar' is merely a disguised description, and, in the last analysis, it is demonstratives that name. On the new view, demonstratives have also acquired an ambiguity. Since even *basic* particulars are complexes, if I say ostensively 'this' in a referential way (I would of course have to say it very fast), I might be referring only to a color in a patch at the center of my visual field (one of the individual "sense data" of earlier theories) or, on the other hand, to the entire complex. (We have here precursors of Quine's rabbits, rabbit parts, and rabbit stages.) Demonstratives, however, do continue to have a crucial role because of the central function of acquaintance with qualities in constructing the ontology of the actual world.

What is important is that ordinary objects and basic particulars are no longer very different kinds of objects; ordinary individuals such as physical objects and persons are certain occurrent series of basic particulars. "A complex of compresence which does not recur takes the place traditionally occupied by 'particulars': a single such complex or a string of such complexes causally connected in a certain way is the kind of object to which it is conventionally appropriate to give a proper name." [10]

Just as we might succeed in referring to a basic event by means of a demonstrative without *knowing* every quality in the complete complex to which we are referring, [11] so we might succeed in referring to an ordinary object without knowing every feature of its history.

All this is best illustrated by Russell's remark about the name 'Caesar' and about demonstrative and pronominal reference:

> [Caesar] was . . . a series of events. . . . If we were to define 'Caesar' by enumerating those events, the crossing of the Rubicon would have to come in our list and "Caesar crossed the Rubicon" would be analytic. But in fact we do not define "Caesar" in this way and we cannot do so, since we do not know all his experiences. What happens in fact is more like this: Certain series of experiences . . . make us call such a series a "person." Every person has a number of characteristics that are peculiar to him; Caesar, for example, had the name "Julius Caesar." Suppose P is some property which has belonged to only one person; then we can say "I give the name 'A' to the person who has the property

10. *Human Knowledge*, p. 308.
11. "I can perceive of a complex of compresent qualities without necessarily perceiving all the constituent qualities." Ibid., p. 302.

P.'' In this case the name ''A'' is an abbreviation for ''the person who had the property P.'' It is obvious that if this person also had the property Q, the statement ''A had the property Q'' is *not* analytic unless Q is an analytic consequence of P.[12]

Having given us this account of ordinary proper names, Russell goes on to say,

> This is all very well as regards an historical character, but how about someone whom I know more intimately, e.g., myself? How about such a statement as ''I am hot''? . . . Here ''I-now'' may be taken as denoting the same complex that is denoted by ''my total present momentary experience.'' But the question remains: How do I know what is denoted by ''I-now''? . . . on no two occasions can the denotation be the same. But clearly the words ''I-now'' have in some sense a constant meaning; they are fixed elements in the language. We cannot say . . . ''I-now'' is a name like ''Julius Caesar'' because to know what it denotes we must know when and by whom it is used. Nor has it any definable conceptual content, for that, equally, would not vary with each occasion when the phrase is used. Exactly the same problems arise in regard to the word ''this.''[13]

He concludes that although ''I-now'' and ''this'' are not ordinary proper names, they have important similarities to such names since both can be used to refer to a larger complete complex of which the constituents *may not be known*.

As we see from the above quotations, some small steps are required for arriving at a direct-reference theory. If Russell had said—which is what I believe would have suited him—that the description of choice in the case of 'Caesar' did not *define* 'P' but was used (employing Kripke's distinction) to fix the referent of the ''tag'' 'Caesar', such a step would have been taken. Similarly 'this' and 'I-now' are again ways of directly referring to basic elements—momentary complete complexes.

But, having failed to take those steps, Russell comes in my view to a mistaken conclusion about which even he is diffident. He says,[14] ''I may perceive a complex without being aware of all its parts . . . I may by attention arrive at a judgment . . . 'P is part of W' where 'W' is a proper name of the perceived complex. . . . But . . . with

12. Ibid., p. 301.
13. Ibid., pp. 301–302.
14. Ibid., p. 303.

better knowledge, our *whole W* can always be described by means of its constituents . . . I think therefore, *though with some hesitation,* there is no theoretical need for proper names as opposed to names of qualities and of relations. Whatever is dated and located is complex and the notion of simple 'particulars' is a mistake." But what does "theoretical" mean here? It is clear, given that we are not omniscient, yet in a public language we can succeed in referring and communicating without any particular shared description in mind to fix the referent of a proper name, that such names *would* be theoretically required. What may be lurking behind Russell's remarks is the held-over belief that, if something has a proper name that named object is not further analyzable; the object itself must be "bare."

The matter can be put in another way. Russell wants to distinguish himself from Leibniz, for whom *all* subject-predicate statements are, in the last analysis, analytic, and to that end he draws a distinction between a judgment of analysis and an analytic judgment. Consider, he says, basic particulars (momentary maximal complexes of compresent qualities). Since a basic particular can be *perceived* without a perception of all its constituents, the namelike 'this', on that occasion, can be given to the entire complex. If it is observed that redness is one of the constituent qualities of the complex, that knowledge is expressed by 'this is red', which, and I quote, "is a judgment of analysis but not in a logical sense an analytic judgment" because "the *whole* was defined as 'this', *not* as a complex of known parts." [15] Then in conclusion he says, "We may agree with Leibniz to this extent, that only our ignorance makes names for complexes necessary. . . . The need for proper names is therefore bound up with our way of acquiring knowledge and would cease if knowledge was complete." But this is an uninteresting conclusion. Acquisition of knowledge is a common enterprise. Knowledge acquired can be communicated and built upon over time. It is an evolving enterprise, and a public language is required for such communication. There is no reason to suppose it will ever be complete. The absence of proper names would thwart such communication and in *that* sense use of proper names *is* theoretically required. It is a condition for making knowledge and communication possible; for getting the enterprise off private ground.

15. Ibid., p. 302.

13. Possibilia and Possible Worlds

The source of this essay was published in *Grazer Philosophische Studien*, 25/26 (1985/86): 107–133, in a double volume on the topic "Non-existence and Predication." It was also presented at the Collège de France, the first of two lectures, in May of 1986; for me a memorable occasion. The present printing contains some stylistic and clarificatory corrections. Some of the exposition was in the interest of accommodating an international audience.

The paper overlaps with an earlier one, "Dispensing with Possibilia," delivered as the Presidential Address at the meeting of the American Philosophical Association, Western Division, on April 30, 1976, and published in the *Proceedings of the Association*, vol. 49, pp. 39–51. That address, omitted from this volume, says more about the uses of substitutional semantics in dealing with contexts containing nonreferring names than does this paper.

I have benefited in this writing from the work, published and unpublished, of Joseph Almog, Keith Donnellan, David Kaplan, Saul Kripke, Leonard Linsky, John Perry, and Howard Wettstein. ∎

Introduction

The development of quantified modal logic[1] has rekindled an interest in systematic accounts of modalities. Modal propositional logic had been widely perceived as, at best, a misleading way of representing metalogical and semantical notions, such as validity, consistency, logical consequence, and the like, through locutions like 'necessity', 'possibility', and 'entailment'.[2] With quantified modal logic (QML) the attack became sharper, for here it was claimed that there emerged deep failures and confusions that went beyond the merely misleading representation of metalogical concepts. The criticism fell into several categories, often insufficiently distinguished.

I will respond to this criticism and discuss two among these continuing debates. The first (to be discussed only briefly) is the putative problem of failure of substitutivity of coreferential terms in modal contexts. The second is the apparent commitment of modal semantics to possible objects, possibilia. In conclusion I will discuss alternatives for a more adequate modal logic and its semantics, developing some themes and proposals from earlier work.

Substitutivity of Identity

The dilemma about substitution has been seen as the following predicament. Either QML leads to a failure of truth-preserving substitutivity for co-referential names and, hence, to the breakdown of the centrally valid principle of substitutivity for identity, or, alternatively and equally odiously, if substitutivity is to be preserved, the values of variables for modal logic will be individual concepts or possible objects rather than ordinary objective referents. On this latter account, substituting 'the evening star' for one occurrence of 'the morning star' in 'Necessarily the morning star is the morning star' will not be permitted, since those terms in modal contexts do not designate the same concept or the same possible object. But then, it is claimed, QML has a different subject matter from that of nonmodal logic, a different

1. See my "A Functional Calculus of First Order Based on Strict Implication," *Journal of Symbolic Logic*, XI (1946): 1–16, and "The Identity of Individuals in a Strict Functional Calculus," *Journal of Symbolic Logic*, XII (1947): 12–15. See also Rudolf Carnap, "Modalities and Quantification," *Journal of Symbolic Logic*, XI (1946): 33–64, and *Meaning and Necessity* (Chicago: University of Chicago Press, 1947).

2. See W. V. Quine, *From a Logical Point of View* (Cambridge: Harvard University Press, 1953). Also, "Three Grades of Modal Involvement," in *The Ways of Paradox* (New York: Random House, 1966).

domain over which its variables range. QML is not seen as a straight-forward extension of standard predicate logic achieved by adding modal operators and syntactical rules for their use along with additional inference rules. The *semantics* shifts radically, from ordinary object domains to domains of concepts [or other intensional objects]. Russell's propositions are replaced by something like Frege's thoughts.

Russell's propositions are structures that have ontologically independent actual objects, not concepts or meanings, among their constituents. Recall Frege's astonishment when he was told by Russell[3] that Mont Blanc, the mountain, was "itself a component in the proposition 'Mont Blanc is 4,000 meters high'." How much confusion would have been avoided if an alternative vocabulary, such as the vocabulary of "states of affairs," had been deployed, avoiding the word 'proposition,' which for many suggests a linguistic or quasi-linguistic entity or a mental or quasi-mental entity.

It was perhaps for such semantical reasons that, of the early modal extensions of quantification theory, mine and then Carnap's[4] (although there were syntactical similarities), Carnap's was set aside. Carnap's background semantics involved *inter alia* systematic ambiguity of reference relative to context in a Fregean way and ran counter to accepted views of objective reference, views that receive a realization in standard nonmodal model-theoretic semantics.

I will not dwell long on the Quinean criticism to the effect that modal extensions of quantification theory, in the interest of preserving substitutivity, *must* run afoul of the effective standard semantics; for, shortly after Quine presented this view, Arthur Smullyan,[5] Frederic Fitch,[6] I, and others[7] showed that analysis in terms of Russell's theory of descriptions dispelled the puzzles Quine had raised; there was no need for a radical shift to Fregeanlike concepts or individual essences or intensional objects as objects of reference.

On the theory of descriptions, the identity sign is, after analysis, *never* flanked by descriptive phrases. A singular descriptive phrase

3. G. Frege, *Philosophical and Mathematical Correspondence,* ed. G. Gabriel et al. (Chicago: University of Chicago Press, 1980), p. 163.

4. *Meaning and Necessity.*

5. A. F. Smullyan, "Modality and Description," *Journal of Symbolic Logic,* XIII (1948): 31–37. See my review of Smullyan, ibid. (1948), reprinted in this volume.

6. "The Problem of the Morning Star and the Evening Star," *Philosophy of Science,* XVI (1949): 131–141.

7. In particular in "A Functional Calculus of First Order Based on Strict Implication," the cases of apparent substitution failure are ruled out as a special case of a general theorem about constraints on substitution in modal contexts. See also "Modalities and Intensional Languages," this volume.

specifies a set of properties that uniquely characterize an object if there is one. It does not refer directly to that object. I need not review the details of the analysis, since they are by now well known. Suffice it to say that the troublesome premises that claim necessity for the identity of the morning star and the morning star and also contingency for the identity of the morning star and the evening star, unpack into complex pairs of premises, given that there are scope ambiguities in the location of the quantifiers and modal operators. But, on either of the alternative readings, a conclusion to one of the analogously unpacked versions of 'Necessarily the evening star is identical to the morning star' that in surface grammar seems to contradict the original premises does not in fact follow at all.[8]

It is a curious fact that Quine, who leaned on the theory of descriptions in ''On What There Is''[9] as a solution to puzzles about nonreferring singular terms, failed to see its effectiveness in dispelling his apparent puzzles about substitutivity in modal contexts.

Possibilia

Rather more interesting is the next locus of criticism: that QML would seem to admit ''possibilia.'' In discussion of this issue it is helpful to summarize how standard semantics has been extended for QML in such a way as to admit possibilia.[10]

Formal semantics in nonmodal model theory for first-order logic, according to what was and continues to be the prevailing practice, *begins* with a given domain of individuals. There are general constraints on what counts as an individual of a seemingly redundant kind. Individuals are distinct from one another, and each is self-identical. They have properties and enter into relations. Individuals can be assembled into sets and ordered into *n*-tuples, which in turn can be assembled into sets. *n*-tuples (ordered sets of individuals) may be viewed as constructed objects extending the domain of objects beyond the initial set of basic individuals. There are also vaguer restrictions. Individuals must have a certain definiteness, a closure, the seeming lack of which in concepts prompted Frege to deny that a concept *was* an

8. See Smullyan, ''Modality and Description,'' and Quine, *From a Logical Point of View* (Harvard, 1953).

9. In *From a Logical Point of View*.

10. My exposition here of the semantics for QML will follow Saul Kripke, ''Semantical Considerations on Modal Logic,'' *Acta Philosophica Fennica, Proceedings of a Colloquium in Modal and Many Valued Logics* (1963), pp. 83–94.

object. It is redundant to say that objects, unconstructed or constructed, are *individuated*. That is at the heart of the slogan "No entity without identity."

Present accounts lack much of the additional metaphysical content present in Russell.[11] Individuals need not be "logical atoms" in a reductionist sense. They need not be metaphysically irreducible. Disagreements about the choice of an object domain in an interpreted theory are thrashed out elsewhere. There are no restrictions on how many individuals there are or how many sorts or categories. But with sorts, there will be sortal modifications for quantification that could take the form of introducing type levels. What is required is that there *be* things that, taken together, make up the domain of the model, its ontology. If the syntax has individual constants, that is, directly referring names of individuals as well as individual variables, the language is interpreted by assigning an object from the domain to an individual name, a truth value to a sentence letter (zero-degree predicate), unordered sets to unary predicates, sets of *n*-tuples (*n* objects ordered) to *n*-ary predicates, and so on. An atomic sentence is true just in case the singular individual or *n*-tuple assigned to the constants in appropriate sequence are in the set assigned to the predicate. A universally quantified sentence '$(x)Fx$' is true in a model M just in case the sentence 'Fa' is true in M and in all variant models like M, except for the object assigned to the individual constant 'a'. An existentially quantified sentence '$(\exists x)Fx$' is true in M just in case 'Fa' is true in M or there is at least one such variant model in which 'Fa' is true.

Given the additional familiar truth conditions on the sentential operators, truth is defined for all the sentences of first-order logic under the given interpretation. Logical truths are defined as those that are true under any interpretation or model. If it is believed, given reductionism and extensionality, that first order logic is sufficient for presenting any "theory," broadly conceived, then, it is claimed, the ontology of a theory may be taken to be the domain of its basic individuals, for it is *only* of those entities that we make, via quantification, existence claims.

We see on this model-theoretic analysis that one cannot make straightforward nonexistence claims such as 'The winged horse does not exist' where 'the winged horse' is taken as a name. As we noted

---

11. *The Problems of Philosophy* (New York: Holt, 1919). "The Philosophy of Logical Atomism," in *Logic and Knowledge*, ed. R. C. Marsh (London: Allen and Unwin, 1956).

above, an analytical device is required in order to give the logical form of such sentences with singular descriptions, of which the theory of descriptions is a most effective example, serving as it does to clarify many philosophical puzzles. Upon analysis the sentence in question reduces roughly to 'There is no one thing that is winged and a horse'. Since no individual satisfies those conditions, nothing nonexistent has been named. If one requires that objects be the objects of this world or grounded in this world (whether or not they are also in alternative structures or worlds), *merely* possible objects are thereby excluded. This requirement is applicable not only to concrete particulars such as Russell's sense data or Carnap's physical objects. Abstract particulars or universals such as numbers, sets, properties, and relations must also be actual, albeit not material. For early Russell, who insisted that we maintain a robust sense of reality, universals are real, although they may be sorted from basic individuals with respect to type. It is redundant to say of real objects that they exist, and the puzzle of possibilia is solved by analysis. This has been a prevailing view about possibilia from Russell through Quine.

But the Kripke extension of model-theoretic semantics to QML[12] *does* seem to admit possible objections that cannot be analyzed away. There are at least four questions to be asked: (1) Does QML in fact admit possibilia? (2) Is it plausible that they be countenanced and admitted? (3) Are there insuperable obstacles to our understanding of their role—can we make sense of them? (4) *Must* QML admit possibilia?

To explore such questions, it is useful to summarize roughly the way in which [Kripke-style] modal semantics extends the standard theory. We have a set of worlds, constructions if you like, including the actual world. Assigned to each is a domain of objects. Domains of worlds may be coextensive (the same set of individuals in each world), intersecting, or disjoint. Where domains are not coextensive with or included in the domain of the actual world, possibilia have been admitted.

A reflexive relation R, between worlds, is introduced where "$W_m R W_n$" corresponds informally to "W_n is possible relative to W_m" or alternatively "W_n is accessible to W_m."

For sentential modal logic, each world is defined by a set of truth assignments to atomic sentences. Truth of all sentences is relativized to worlds. Where the major connective of (or operator on) a sentence is truth-functional, its truth value in a world W_m is a function of the

---

12. Kripke, "Semantical Considerations on Modal Logic."

truth of its parts in W_m. Where the major operator is modal, as in '$\Diamond p$' and '$\Box p$', its truth in W_m is a function of the truth of 'p' in relatively possible (accessible) worlds. '$\Box p$' is true in W_m just in case 'p' is true in all worlds W_n such that $W_m R W_n$. The truth of '$\Diamond p$' requires at least one accessible world W_n where 'p' is true; this is reflected in the axiom '$p \supset \Diamond p$'.

Familiar additional candidates for modal axioms such as

$$\Box p \supset \Box\Box p$$

$$p \supset \Box \Diamond p$$

reflect additional constraints on R—in case of the above, transitivity and symmetry, respectively. Taken together, these axioms guarantee that R is an equivalence relation.

For sentential modal logic, the truth definition is a natural extension of the truth definition for nonmodal logic. Each world is defined by a set of truth-value assignments to sentence letters. Given the usual truth rules for the truth-functional connectives, the assignment of truth values to compound sentences in *each* world is recursive on the initial assignments in those worlds. It remains only to extend the truth definition to modal sentences to arrive at the semantics for *modal* sentential logic. A sentence '$\Box p$' is true in a world W_i just in case it is true in all worlds to which it is accessible. Validity of a sentence is defined as truth in every model in every model structure.

Questions about possibilia arise when the semantics is extended to quantification theory. In my own sketch of a semantics for modal logic,[13] the domains of individuals assigned to alternative worlds were *coextensive*. Given that one of the worlds is the actual world, no entities are spawned that are not in this world and no entities of this world are absent in others. That is consistent with the axiom, variations on which came to be known as the "Barcan formula," which said: If $\Diamond(\exists x)\varphi$, then $(\exists x)\Diamond\varphi$, and, derivatively, it follows that $\Diamond(\exists x)\varphi$ if and only if $(\exists x)\Diamond\varphi$.

Since we have the option of coextensive domains, QML is not *committed* to possibilia. Yet admission of possibilia would *seem* to be a natural extension, for informally, the notion of possible worlds lends itself to framing counterfactuals not merely about the properties actual objects might have and the relations into which they might have entered but about alternative worlds that might have individuals that

---

13. "Modalities and Intensional Languages," this volume, concludes with a sketch of a semantical construction where domains are coextensive, which underpins the Barcan formula as formally proposed in the 1946 and 1947 articles in the *Journal of Symbolic Logic* cited in note 1 above.

fail actually to exist, or fail to have individuals that do actually exist. The semantics accommodated such interpretations.

In setting up his quantificational structures, Saul Kripke singled out the actual world as worthy of special designation among alternatives, but in the continuing literature special mention of the actual world did no work and was usually dropped. In the course of this discussion I will argue that the special role of the actual world is crucial for addressing the question of possibilia.

On Kripke's quantificational syntax and semantics for QML there are no individual constants. Predicate letters are extensionally defined; an n-adic predicate letter is assigned a set in *each* world; a set of ordered n-tuples for $n > 1$. Formulas consisting of predicate letters followed by variables become atomic sentences in a world W_i on the assignment of objects to the variables from the joint domains of all accessible worlds.

There is then the problem of assigning a truth value to a sentence about an object in a world where that object fails to exist. Suppose, Kripke speculates, Sherlock Holmes exists in some nonactual world. What truth value should be assigned to the sentence 'Sherlock Holmes is bald' in each world? If we deny a truth value to those atomic statements in worlds where those objects are absent, we must have a theory with truth-value gaps, akin to a view of Frege or Strawson.[14] Russell, who took such names to be abbreviated descriptions, would always assign a value to such a sentence. The sentence 'Sherlock Holmes is bald' would, for example, be false for Russell, whose view of truth was centered in the actual world. Kripke goes on to say that our choice in such cases is a matter of convention. Although on Kripke's semantics the sets assigned to predicates in a given world contain objects only from the domain of that world, he adopts the ''convention'' that all atomic sentences will have a truth value, even in worlds where the objects assigned to the variables in atomic sentences are not in the domain of that world. (Variables do double duty on Kripke semantics: they may serve as individual constants as needed.)

But *quantification*—universal and existential generalization and instantiation—*is* world-bound. '$(x)Fx$' will be true in a world W_i just in case 'F_iy' is true for any assignment to y from *the domain of W_i*. For *quantificational* purposes the variables range over only the domain of that world. For *naming* purposes they range over domains of all

14. G. Frege, ''Sense and Nomination,'' in *The Philosophical Writings of Gottlob Frege,* trans. Peter Geach and Max Black (Oxford: Blackwell, 1952); P. F. Strawson, ''On Referring,'' *Mind,* LIX (1950): 320–344.

worlds. It can then be shown in such a semantics that the Barcan formula and its converse fails. Neither '$\Diamond(\exists x)\varphi x \supset (\exists x)\Diamond\varphi x$' nor its converse is valid.

Despite the elegant generality of the formal extension of model theory to modalities that allows variable domains, many of the examples given of possibilia are among those that could be accommodated by the theory of descriptions and do not seem to require variable domains. Nevertheless, the semantics *accommodates* possibilia. But modalities in their primary use concern counterfactuals about actual objects, and to reintroduce possibilia is to run counter to the admonition of Russell that we "retain our robust sense of reality." Why spawn possibilia again if they can be analyzed away? The issue is, *Can* they be analyzed away in modal discourse?

Putative possibilia are familiar artifices; we concatenate some set of predicates attached to a uniqueness condition and endow it with reference to some kind of shadowy object. But does a mere concatenation of properties make an object? The Russellian strategy is to employ description theory, which permits us to say what we want to "about" winged horses and the like without there having to *be* any.

But some would point to more plausible candidates for possibilia. Surely *some* names of nonactual objects seem to be more than contrived Meinongian recipes for spawning objects. They are names like 'Apollo', or 'Pegasus', or 'Père Noel', or perhaps 'Atlantis' or 'the river Styx', which have a richness, vividness, and entrenched continuing history in the shared linguistic institution and in shared beliefs over time.[15] But it is surely not the richness of association and shared thoughts that grounds reference to actual objects. When I say that Durandus was a thirteenth-century Dominican philosopher, that may be all I know about him; not even a *unique* set of properties may be available, to say nothing of a rich narrative or picture. Still I *am* referring to an actual object. If richness of association is neither a necessary nor a sufficient condition for reference to an actual object, on what grounds is it a basis for positing a possible object?

There are of course the following kinds of claims that suggest a plausible reason for entertaining possibilia: there might have been more things than there are; there might have been different things than there are.

---

15. David Kaplan in "Quantifying In," in *Words and Objections,* ed. D. Davidson and J. Hintikka (Dordrecht: Reidel, 1969), pp. 178–214, stresses the richness of a concept as a possible condition for its serving as a "standard name."

Problems about Possibilia

I want to review some of the Quinean[16] arguments against possibilia because they articulate what has gone wrong in adopting some versions of modal semantics, such as Kripke's in its original presentation. Quine, who sees himself as a natural heir of Russell, is guided by some but not all of Russell's methodological norms. In "The Philosophy of Logical Atomism" Russell describes his twofold philosophical purpose as, first, advancing the view that one can arrive at ultimate simples, where included among simples are particulars as well as qualities and relations, and, second, what is characterized as the application of Occam's razor. The task, he says, is to find "the smallest *apparatus,* . . partially but not necessarily wholly empirical out of which all propositions can be constructed." Those Russellian ends are reflected in the architectonic of *Principia Mathematica*. The post-Russellian logic and the evolution of formal semantics in Carnap, Tarski, and thereafter were also reality (actuality) motivated. They were informed by Russell's twofold purpose, but with some modification. There is on later views greater neutrality about the choice of a domain of particulars or individuals. In the later theories, Russell's type-theoretic structure, which includes a panoply of qualities and relations, is "reduced" by principles of extensionality to sets of individuals or sets of ordered *n*-tuples (ordered structures) of individuals. Thus such higher-order "simples" as Russell's qualities and relations are taken to be constructions out of basic particulars. That reduction is perceived as a further advance in the program of finding "the smallest apparatus not necessarily wholly empirical out of which propositions can be constructed."

On the level of individuals Quine shows extraordinary flexibility with respect to rival ontologies. There are no privileged domains of particulars beyond the empiricist norm that he shares with Russell: that a theory must at least be able to accommodate the physical objects and propositions of science broadly conceived. Object domains among rival theories may vary with respect to aesthetic or metaphysical appeal; each may be fundamental in its fashion, and we are exhorted to be tolerant.

Now, given the exhortation to be tolerant and the acceptability of either phenomenalism with its ontology of sensory events or physicalism with its ontology of physical objects or events as equally viable alternatives among others, it is difficult to appreciate initially the rid-

16. Quine, "On What There Is," *Review of Metaphysics,* 2 (1948): 21–38. Also in *From a Logical Point of View*.

icule that Quine heaped on the philosopher he calls "Wyman," who, it seems, was foolish enough to admit *possibilia* into his ontology.[17] Quine distinguishes two grounds for rejecting possibilia. One is Occam's razor. Possibilia overpopulate the universe. But, as it turns out, that particular application of Occam's razor is confessed to be aesthetic, and Quine admits a taste for desert landscapes. I have never appreciated that use of Occam's razor if it was intended that we take it seriously. Plenitude is equally compelling as an aesthetic category, and, as Leibniz would have us believe, it was God's preference. If nature abhors a vacuum, it may prefer a plethora, Furthermore, if it is a question of overpopulation, how does one take an ontological census? The physicalist who admits a few possibilia may well be countenancing far *fewer* entities than the phenomenalist with all of those subjective events of sensation it takes to construct even one physical object.

But the rejection of possibilia is not claimed on aesthetic grounds alone. A more serious ground is the claim that possibilia "cannot meaningfully be said to be identical with themselves and distinct from one another."[18] If that is so, then of course possibilia do not meet minimal conditions for being individuals.

Quine does not attempt to defend his claim as originally stated. Rather he adopts some *identifiability* thesis where the meaningfulness of an identity sentence '$a = b$' *always* requires that we have *criteria* for deciding that a and b are the same thing. The slogan is now "No entity without *identifiability.*" Let us confine ourselves in the following discussion to a consideration of material or physical objects and persons, for these are familiar cases. There are, of course, interesting questions about identification of *abstract* individual objects, such as numbers, or higher-type abstract empirical kinds, such as gold or elm, which can be shown to illuminate problems about the identification of material or physical particulars. But at this juncture our attention will be directed toward nonabstract individual physical objects or persons that have been taken to be the least controversial candidates for the successful specification of criteria of identification, criteria that would provide necessary and sufficient conditions for counting such an object a and such an object b as the same object.

The rejection of possibilia on the ground that *criteria* of identification are unavailable would be partially justifiable if it could be shown that such criteria could *always* be given for actual objects. But

---

17. "On What There Is."
18. *From a Logical Point of View*, p. 4.

that too is at issue. It has been argued, persuasively, that efforts to provide such criteria *universally and necessarily* applicable to actual objects may also not be achievable.[19] Failure to locate criteria of identification would therefore not distinguish actual objects from possible objects. Of course one might arrive at sufficient conditions, such as having the same continuous spatiotemporal path, as in the case of the identification of Hesperus and Phosphorus as a single planet, but such continuity may not be necessary, as in versions of the puzzle of Theseus's boat. That there might be necessary conditions, sameness of some qualitative properties, that fail to be sufficient for identification, does not surprise us. But there are grounds for saying that even *indiscernibility* in some far stronger sense may not be sufficient for guaranteeing identity.

Identity and Indiscernibility

Identity is the strongest equivalence relation that a thing bears only to itself. That there are individuals is already presupposed if the identity relation is to hold. The identity relation does not *confer* thinghood; identity is an essential feature of things. Individuals must be there before they enter into any relations, even relations of self-identity. Of course if we want to *discover* which objects a language or theory takes to be individuals, we look to see which objects are such that they can meaningfully enter into the identity relation. Quantification is not so clear a guide to ontology as is identity.[20] No identity without entity.

There are of course variant versions of principles that tie indiscernibility to identity, and I would like to discuss some of them, since they may provide some insight into the ultimate failure of all searches—which are often undertaken—for identification conditions given in terms of abstract, nonindexical sets of properties.

Indiscernibility has metaphysical and epistemological versions, and under each classification there are multiple views of what counts as having all properties in common. There is the narrow metaphysical version, presumably Leibniz's, that requires that indiscernible entities have only *qualitative* properties in common. These exclude spatiotemporal properties and relations and other nonqualitative "external" properties and relations.

19. See Eli Hirsch, *The Concept of Identity* (New York: Oxford University Press, 1982). Also, S. Kripke, "Second Discussion," *Synthese*, XXVII (1974).
20. See "Modalities and Intensional Languages," this volume.

But here there is no guarantee that an *a* and *b* with all such qualitative properties in common will be *numerically* the same. If numerical identity is required for identity, indiscernibility fails as a guarantee. For Leibniz, God and a metaphysical principle about sufficient reason were required for indiscernibility to entail identity in this more narrow of the metaphysical senses. F. P. Ramsey,[21] in questioning Russell's "definition" of identity in terms of some noncircular indiscernibility condition, argues that it is not self-contradictory for two things to have all their elementary qualitative properties in common. Although identity entails narrow metaphysical indiscernibility, the converse holds only on some additional assumptions, as Leibniz clearly saw. If the whole gamut of shared elementary qualities do not give us sufficient conditions for metaphysical identity without additional metaphysical assumptions, why should one expect that some more limited set of properties could be found for special cases like material objects or persons that could *guarantee* identifiability in all cases?

One can of course insist on taking indiscernibility in this narrower metaphysical sense as entailing identity, but then one is left with the bizarre conclusion that numerically distinct objects might be identical. When Russell[22] in his later work adopted a Leibnizian view of particulars as internally structured maximal conjunctions of qualities, he seemed to accept the bizarre consequence that if there were a tower in New York that was the same complex of compresent qualities as the Eiffel Tower in Paris, then they would be one and the same. He comforts us with the claim that, on empirical inductive grounds, such a coincidence is highly improbable.

Of course if metaphysical indiscernibility is *so broadly* defined that any partially interpreted sentence with a free variable, i.e., any propositional function, defines a property, then of course indiscernibility will entail identity, since if *a* and *b* are indiscernible in this global way, then each has the property of being identical with *a* and being identical with *b,* and so *a* and *b* are identical *simpliciter.*

We can derive identity from indiscernibility on this wide view only because identity is already given to us as a primitive relation. Among the properties we take *a* and *b* to share when identical are the very properties of being identical to *a* and of being identical to *b.* The definiendum is already included in the definiens, for identity is the

21. *The Foundations of Mathematics* (New York: Humanities Press, 1950), p. 31.
22. *An Inquiry into Meaning and Truth* (London: Allen and Unwin, 1950), pp. 96–97.

strongest of the equivalence relations that are transitive, reflexive, and symmetrical. If our formal languages are expanded to include modalities, then being necessarily identical to *a* and being necessarily identical to *b* will also be properties on the unrestricted metaphysical version of identity, and of course the necessity of an identity will follow.[23]

We have seen that metaphysical indiscernibility, narrowly conceived as the sharing of ''all'' properties but with a restrictive account of which properties count for indiscernibility, is no guarantee of identity without some strong metaphysical assumptions. Alternatively, on the unrestricted account that identifies any function with a property, indiscernibility already presumes identity, and the definition is circular. I am not suggesting that the circularity is vicious.

For epistemological indiscernibility of an object *a* and an object *b,* the requirement is not just that *a* and *b* have all their properties in common but that there be some way for the knowing subject, and the epistemological agent, to make that determination. A more strongly verificationist claim is that if a property or relation cannot be known to an agent, then it does not count as a property or relation at all. The search for criteria of identity is grounded in the belief that one can *know* some shared set of properties of an object *a* and an object *b* that guarantees their identity, and, on verificationism, those are the *only* properties that count at all. Indeed, the term 'indiscernibility' already comes with strong epistemological overtones, as in the shift in meaning from 'different' to 'discernibly different'. The term suggests a strong connection between a metaphysical relation, identity, and the epistemological agent's capacity to know or *discern* sameness and difference—a Kantian as well as a verificationist theme.

It would strengthen the tie between identity and indiscernibility in the epistemological sense if one could show that where a difference in properties between an *a* and *b* cannot be *known* to an epistemological agent, *a* and *b* must be identical; or, more strongly, if one could show that there is no effective use for the distinction between metaphysical and epistemological discernibility. But that is difficult to maintain. There is an interesting passage in Carnap's *Der Logische Aufbau der Welt*[24] worth summarizing. Consider, he says, the following: We are given a map of the Eurasian railroad with the usual

23. See ''The Identity of Individuals in a Strict Functional Calculus of First Order'' (1947), and ''Modality and Intensional Languages'' (1961), this volume, for proofs and discussion of the necessity of identity.
24. *The Logical Structure of the World,* trans. R. George (Berkeley and Los Angeles: University of California Press, 1967), pp. 25–27.

distortions of ordinary maps and a topological rather than a metrical representation of its properties such as connections and intersections. It marks stations as points, but gives no names or entries other than the railroad lines. Can we determine the names of the stations through inspection of the actual network? Carnap tells us how to proceed, but concludes that "if there should still be two locations for which we have found no difference even after exhausting all available scientific relations, then they are indistinguishable, not only for geography but science in general. They may be *subjectively* different; I could be in one of those locations but not in the other. But this would not amount to an objective difference, since there would be in the other place a man just like myself who says as I do: I am here but not there." Since Carnap *defines* objectivity in terms of epistemological discernibility, he says that the difference in the *indiscernibly* different locations is not objective. Still, there are two locations.

Carnap's example, like the symmetrical mirror-image universes, discloses that the *ultimate* determination of the truth of an identity statement cannot be guaranteed by description in terms of nonindexical, nonostensive properties and relations alone. It requires ostension at some juncture. If one views proper names as meaning-free, as tags for pure reference, then they too serve a quasi-indexical role, but with an advantage. On a causal or historical-chain theory of direct reference,[25] proper names have an ostensive origin, are theoretically traceable causally or historically, and are part of a publicly available vocabulary of names with fixed reference in the language. Proper names serve as a long finger of ostension over time and place.

On this account, "proper name" is a semantical, not a merely syntactical, notion. Reference is supposed. We may mistakenly believe of some syntactically proper name, say 'Homer', that it has an actual singular referent and is a genuine proper name, but if its use does not finally link it to a singular object, it is not a genuine name at all.

It is for such reasons that linguists exclude proper names from the lexicon altogether.[26] *Qua* proper names, they don't have meanings as ordinarily viewed. In the case of names of persons for example, biographical dictionaries, "dictionaries" of proper names, help, through

---

25. See K. Donnellan, "Reference and Definite Descriptions," *Philosophical Review*, LXXV (1966): 281–304; "Proper Names and Identifying Descriptions," *Synthese* (1970). See also S. Kripke, "Naming and Necessity," in *Semantics of Natural Language*, ed. D. Davidson and G. Harman (Dordrecht: Reidel, 1972).

26. P. Ziff, *Semantic Analysis* (Ithaca: Cornell University Press, 1960) and elsewhere, adopts the linguists' view. Proper names are not "part of the language" in the sense of being lexical items in a dictionary.

203

description, to tell us what or who is being referred to, but that is not given as the "sense." The referent of a proper name remains fixed, even where attributions claimed by narrative are in error.

Proper names have fixed values in our language as a historical institution and are a part of the public vocabulary. In this way they allow reference to an object despite the vicissitudes the objects undergo and despite the absence of direct acquaintance with many and perhaps most of the objects that the language user correctly names. What is important in reference to objects like material objects is that there be an episode, *publicly recognizable,* in which an object of acquaintance is named and the name launched[27] into the language. This brief excursion into naming has served as a preliminary to a consideration of *possibilia.* It is not, I will propose, the general absence of "identification *conditions*" that makes possibilia problematic. It is that possibilia cannot be objects of reference at all.

I should now like to get at the core of justification in Quine's rejection of possibilia understood as definite individuals. One must first recall that the question of possible objects may be wholly detached from the modalities of modal logic. In the syntax of modal logic, modal operators attach to sentential functions or sentences, not to names or descriptions. In this respect they are like the quantificational operators. On standard quantification theory, 'Socrates exists' has no well-formed isomorphic syntactical counterpart. In modal quantification theory the same holds for 'Homer is possible'.

Nor are we addressing the question of possibilia as a way of separating individuals of different categories as between empirical and nonempirical. The familiar Meinongian possibilia, the golden mountain and the winged horse, are in the category of material objects. Rather, as in Meinong, it is their metaphysical status that is being distinguished, their different modes of being. Albeit material objects, the Meinongian claims that they may "exist" in different ways, actually, possibly, even impossibly. But these are not the modalities of modal logic. Carnap's possible worlds have, in extensional contexts, coextensive domains of actual objects or, in intensional contexts, coextensive domains of associated concepts. My initial theories had coextensive domains of actual objects in and out of modal contexts.

The central truth in Quine's critique of possibilia is obscured because his antipathy to modal locutions *tout court* blurs such distinctions. But, more important, his criticism is somewhat out of focus

27. See note 25 above.

because it is grounded in an inadequate semantical theory, a theory that claims that the syntactical category of proper names can, in giving the logical form of sentences, be wholly eliminated. Individual variables are, he says,[28] sufficient for bearing the entire burden of reference. "Whatever we say with the help of names can be said in a language which shuns names altogether." In this he follows a disappointed Russell who finally abandoned ordinary proper names as candidates for logically proper names, albeit reluctantly—reluctantly because ordinary names such as 'Scott', 'Napoleon', those ordinary names he used to explicate the theory of descriptions, seemed to retain so many of the appropriate features for vehicles of ostensive reference. But since, for Russell, ostensive reference required acquaintance with the object named by all who use a proper name, it would follow as a consequence that, for those who *have* such acquaintance, proper names have a wholly different semantical role from their role in the language of those who lack such immediate acquaintance. Russell reasoned that in a public language a sentence like "The Eiffel Tower is a steel structure" should not shift in objective content so dramatically as between those who have confronted it and those who have not. So, reluctantly, ordinary proper names were analyzed away as disguised descriptions. The flourish that Quine added, recognizing as Russell did the difficulty of arriving at some favored description as the definiens of a name, was the procrustean solution of converting names like 'Socrates' into predicates like 'Socratize'.

There are features as well as consequences of this device, which suggests legerdemain; as if someone's having the property of being a Socratizer or a Pegasizer were just another property like being a philosopher or a winged horse. But I need not discuss such peculiarities here. The issue is rather that, to get our language off the ground, there must be publicly accessible objects as well as devices for direct reference independent of description. Ordinary proper names as well as other categories of terms such as indexicals constitute such devices. We are not ignoring the fact that we have a limited number of names in our vocabulary. The point is theoretical and philosophical. Actual objects are there to be referred to. Possibilia are not.

In allowing domains of possible worlds to include nonactual objects Kripke[29] allows for possibilia. He says, "We must associate with each world a domain of individuals, the individuals that exist in that

---

28. "On What There Is," pp. 12–13.
29. "Semantical Considerations on Modal Logic."

world. . . . In worlds other than the real one, some actually existing individuals may be absent, while new individuals like Pegasus may appear.''

The semantics for Kripke's theory appears to be symmetric as between referring to actual objects and referring to possible objects. According to the procedure adopted, variables serve the dual role of varying over the world-bound domain for purposes of quantification and serving as individual constants to which objects from any domain are assigned. No special problem is noted about assigning possible objects to individual variables serving as individual constants. What Quine sees is that there *is* such a problem—but his examples are ambiguous, and the problematic cases are not clearly distinguished from the unproblematic cases. Quine says,[30] "Take for, instance, the possible fat man in that doorway; and, again, the possible bald man in that doorway. Are they the same possible man or two possible men? How do we decide? . . . By a Fregean therapy of individual concepts, some effort might be made at rehabilitation; but I feel we'd do better simply to clear Wyman's slum and be done with it." Now in this example Quine is pointing to an actual doorway; that doorway. Is there some actual thin, nonbald man in the doorway, and is the question about whether that actual thin nonbald man might possibly have been fat and bald? But those are familiar, meaningful counterfactual queries about that actual man; they are not about possible objects. If a question is being raised about some actual fat man possibly being in that doorway and some actual bald man possibly being in that doorway, then that question is also about actual objects and not special to possibilia. So the question must be about some nonactual man who *is* in that actual doorway, or about a nonactual man who might be in that doorway. The latter is an even more complex modal question. One can recognize here ontological issues debated by medievals, such as haecceitism.

Less ambiguous examples of the puzzle are helpful. Suppose I say of a given terrain, ''There might have been a mountain here.'' I might even purport to give it a name, 'Mt. White'. My speculation might be reasonable, where the locale has a history of earthquakes and eruptions of a kind that sometimes lead to such formations. Suppose there is then an eruption and a mountain forms. Could I claim that a possible individual, Mt. White, has become actual? Of course not. To be a material object the object must have had a unique and traceable history in a material order of things. It isn't a thing waiting in the wings to

30. "On What There Is," p. 4.

take its place among the actuals when called. There was at the time of "naming" no history of a definite, albeit only possible, mountain, such that the very possible mountain that I claimed to name 'Mt. White' was propelled onto the stage of the actual world. This is rather a case of a thing in the actual world having its inception at a particular time and place. Only then is it properly nameable.

Consider a more familiar case. Certain astronomical perturbations lead to a hypothesis that there is a planet in a certain location. I might even purport to name it. Such was the case in the discovery of the planet Neptune. Can I not view my hypothesis as referring to a possible planet identical to an actual one? If there hadn't been a planet there, if there had been several bodies with their center of mass at that location, would I then just have referred to a possible but non-actual object? Could I say, on discovering Neptune, that the possible planet I referred to is now actual? There is something amiss with that way of speaking.[31] Until Neptune was discovered, that use of the proper name 'Neptune' was not a referring use. Perhaps Leverrier said, "If there is a planet in that location I will name it 'Neptune'." Neptune is at that juncture not named. In Russellian terms Neptune is not a constituent of the proposition nor is the statement *about* Neptune until after the fact. Whatever Leverrier's original intentions, he might on discovery have named the planet 'Ceres'. Of course the *expression* 'Neptune' is a constituent of the proposition about Leverrier's intention to name. It seems harmless in formal semantics to speak of assigning an object from this or any other world to a variable or to a name. But we are in this actual world, users of our actual language. The object must be given in actuality for something like ostension to occur and for the name to refer. Proper naming as opposed to describing defines a special basic relation between a word and a thing in a linguistic institution. Naming relates a word introduced into an actual language in the actual world to a thing that is there to be encountered in the world when the event of naming occurs. Acts of naming are acts of actual language users. A possible object is not there to be assigned a name.

That one has no general criteria of identity for possibilia is not sufficient for rejecting them. As we noted, even general criteria of identification for actual material objects seem, also, to elude us. For an actual *a* and an actual *b* we can refer to *a* and refer to *b* and be in doubt as to whether '*a*' and '*b*' name the same thing. But, in the case of possibilia there is no referring at all. It is not the absence of

31. See "Modalities and Intensional Languages," this volume.

criteria that makes us dubious. It is rather that what is absent is the individuals. They are not there to be objects of reference at all. It isn't that we cannot *settle* the question for a nonactual *a* and a nonactual *b* whether *a* and *b* are identical. There is no plausible question. There are no individual objects, which are what is required for an identity relation. There are no traceable histories, origins, futures, and so on. Criteria aside, Quine is correct when he says, "The concept of identity is inapplicable to 'unactualized possibles'." No identity, no entity.

Most of the familiar Meinongian examples of possibilia effectively fall under the theory of descriptions. The mountain that might have been formed, the planet that might have been in a location, both have a certain plausibility, but for the very reason that a hypothesis about such possibilities has its ground in the actual world. There are other similar cases. The alternative moves in an actual chess game that I did not finally make; a possible outcome of an experiment in process; the house that would have been there had the structure been completed; these are so closely linked to the actual world as to seem reasonable candidates for ostension. But, close as they come, we do not finally confront them. "They" have a past but no present or future in the actual order. In the article "Possibility," Max Black[32] notes the force of such examples and concludes, "There are powerful motives . . . for saying that (such) possibilities are not fictions but in some sense 'real' or 'objective'. But then we want to know *where* they can be— we are embarrassed to find no place for them."

There is something to be learned from exploring the background of the semantical view that allows domains of possible worlds to have nonactual objects. This view gives every appearance of claiming that we can refer to possibilia; such reference would take the form in the structure of assigning possible objects to variables used as individual constants, much as we "assign" to the name 'Socrates' the individual Socrates in a case of actual reference.

In "Naming and Necessity" Kripke[33] musters arguments against a dominant view that ordinary proper names can be eliminated and replaced by singular descriptions, or clusters of descriptions or favored descriptions. Many of the arguments are like those of Russell, who, although he ultimately "reduces" ordinary proper names to disguised descriptions, continues to use the contrast between ordinary names and descriptions to dispel puzzles about nonexistent objects and to explain

32. *Journal of Philosophy*, LVII, 4 (February 1960): 117–126.
33. *Semantics of Natural Language*.

apparent substitution failures in talk about nonexistent objects in contexts of propositional attitudes.

Russell did not specifically address substitution failures in modal contexts, and, although the theory of descriptions works smoothly where it is applicable, in contexts generated by modalities two ordinary proper names of the same thing can always be intersubstituted. For example, truth will be preserved in going from '□(Tully = Tully)' to '□(Tully = Cicero)', whereas truth would seem not to be preserved in going from 'John knows that Tully is Tully' to 'John knows that Tully is Cicero'. Russell gave us many arguments and analytical tools that endorsed a special semantical role for proper names, but then felt he had to concede non-name status even to ordinary proper names. However, early Russell never wholeheartedly abandoned the notion of a *genuine* proper name: an expression with no content or sense that simply refers. But, as we noted, he claimed grounds for denying that role to ordinary proper names. One such ground was his belief that only through direct acquaintance could we name an object with a sense-free name. Another ground would be substitution failures in contexts of propositional attitudes. There is a third ground, and a quotation may be helpful here. Russell says:[34] "If '*a*' is a name it *must* name something; what does not name anything is not a name; and therefore if intended to be a name, is a symbol devoid of meaning, whereas a description like 'the present king of France' does not become incapable of occurring significantly merely on the ground that it describes nothing. . . . And so when we ask whether Homer existed, we are using the word 'Homer' as an abbreviated description; we may replace it by (say) 'the author of the Iliad and the Odyssey'." The intimation here is that, since a genuine name's only "meaning" *is* its referent, it makes no sense to ask whether the name's referent exists, and to entertain this as a meaningful question is at the same time to deny that the "name" is genuine. To restore ordinary names to genuine-name status, a theory had to be provided whereby it could be determined whether a syntactically proper name *was* a genuine name. That was one of the motivations for the causal, or historical-chain, account of direct reference for proper names.

Russell's claim is that, if a proper name is *intended* as a name, it must be supposed that it refers. Hence if it doesn't name, the symbol is devoid of content and must be replaced by a description. But that

---

34. *Introduction to Mathematical Philosophy* (London: Allen and Unwin, 1938), pp. 178–179.

is a radical solution. What is needed is an account of how a speaker's intention to name can be *thwarted,* which can be incorporated into an account of genuine naming. In genuine naming, an object must be given in a public way prior to naming. The name is launched into the language with fixed reference. Once launched, the name becomes part of a historical or causal chain, a chain of communication over time. It is a part of the language as a historical or public institution. A name acquired by a speaker is a genuine name if it can be traced back to the object named. But it can happen that a speaker *intends* to name an actual object and believes he is doing so, but the chain of communication by which he acquired the name does not terminate in an object given and named.

Keith Donnellan,[35] an early proponent of the causal theory, says in "Speaking about Nothing," "Discourse about what *is* (about the actual world) carries the presupposition that the speaker is talking about people, places or things that occur in the history of the world. But sometimes a speaker uses a singular name intending it to refer but it doesn't." He uses the example of *The Horn Papers* published as a diary of one Jacob Horn, who purportedly lived in Colonial America. In fact, *The Horn Papers* were a hoax. The syntactical name 'Jacob Horn' did not terminate in an object so named. It does not succeed in referring. The syntactical name is an invention of the perpetrator of the fraud.

Russell's solution does not help us here. Russell insists that the very questioning of the existence of x where 'x' is a syntactical name *mandates* replacement of 'x' by a description. In the case of the *Horn Papers,* what is a suitable description? Suppose the description is "the author of the Horn papers." That will be satisfied by the *hoaxster;* that author does exist. Furthermore, I need not have any description in mind when I *intend* to refer to an actual person named Jacob Horn. I may claim that Jacob Horn was a Colonial American who wrote a diary, since published. There are many Colonial Americans who satisfy that description. If naming may be viewed as the long finger of ostension over time, the case of a syntactical nonreferring name is like pointing at nothing.

Taking 'Jacob Horn' as nondescriptive, then, on a Russellian view of propositions, the sentence "Jacob Horn was an author of a published diary" does not correspond to a full-blooded proposition. It lacks a constituent. There is no Jacob Horn to be a constituent. There

---

35. *Philosophical Review,* LXXXIII, 1 (1974): 3–30.

is no *x* such that *x* was named 'Jacob Horn', assuming that the syntactical name is univocal.

Peter Geach,[36] an early proponent of this view of proper names, sums it up as follows. "For the use of a word as a *proper* name, there must be in the first instance someone acquainted with the object. . . . But language is an institution . . . and the use of a name for a *given* object like other features of the language . . . can be handed on from one generation to another. Plato knew Socrates, Aristotle knew Plato, Theophrastus knew Aristotle and so on in apostolic succession to our own times."

There are of course the further, more familiar arguments for the failure of the description theory of names, i.e. the theory that names are disguised descriptions. Given that modal discourse accommodates counterfactual claims, the description theory of names also fails. Suppose Homer is a genuine name. Since there are plausible circumstances that might have prevented poets from writing their poems, we can say of Homer that he might not have been the author of the Odyssey and the Iliad; on Russell's definition of the name 'Homer', i.e., 'The author of the Odyssey and the Iliad', that would be a contradiction. One may not be able, without circularity, to generate some favored set of universal essential properties that describes Homer and only Homer. The property of being the thing identical to Homer or the thing that Homerizes will not do. Such devices do not *eliminate* the name; they recycle it. But among the most interesting and decisive of Kripke's arguments is that we correctly use names, perhaps more often than not, without having *any* unique description in mind. Someone might assert, truly, that Homer was a Greek poet. That is all he knows about Homer, yet if the claim terminates properly, he *is* referring to Homer and he quite understands what he is saying.

In addressing the question of possible objects, I have modified some themes in Russell. In particular I share his realistic view of propositions as being not wholly linguistic or quasi-linguistic entities but rather structures with some worldly constituents. The departure from Russell's view is that, as I see it, ordinary individuals as well as properties and relations may be constituents. This realistic account for atomic sentences as representing such structures is relatively unproblematic. Russell's account of variables as also being constituents is more problematic, but need not concern us here.

---

36. *Logic Matters* (Berkeley and Los Angeles: University of California Press, 1972), p. 155.

A feature of genuine directly referring names, in contrast to many descriptions, is that the values of such names remain fixed. But there are also some descriptions that are satisfied across all possible circumstances. If it is claimed that having a particular and uniquely describable biological origin is an essential feature of an organism, that *description* would be *satisfied* by the same individual across possible circumstances. But that description is not a directly referring name. When proper names and such essential descriptions were conflated in a category of rigid designators, a crucial distinction was obscured.[37] Objects must *be there* to be *named*. Descriptions are prescriptions for finding an object. The *relations* between a name and its referent and between a description and its satisfier are different.

Modified Modal Semantics

If the domains of possible worlds, one of which is this world, are coextensive or are subsets of the actual domain, the variables will range over actual objects in all worlds. Quantification which is world bound will be objectual. No perplexities about possibilia.

An object may be assigned by some naming practice to more than one name, but once assigned, the referent or value of a name remains fixed. Names do not specify conditions of satisfaction. They are simple and without meaningful syntactic structure. Speaking slightly paradoxically, the ''meaning'' of a name *is* its referent. It is the feature that makes a name a nonlexical item.

Given the fixed assignments to names, ''objects'' identical in a given world will be identical in all worlds where ''they'' exist. Since atomic sentences may be assigned truth values even where their individual constituents do not exist, '$a = a$' in a world where a is absent will not ground local quantification. Nevertheless the relation of identity is well defined, and there are no contingent identities.

A more radical alternative[38] for modification of modal semantics is to take the quantifiers as substitutional. No domains are assigned to worlds at all; the variables do not range over objects, they are place markers for *syntactically* proper names. Atomic sentences on such an account do not represent structures with constituents. They represent unstructured ''contents.'' Truth or falsity may be assigned to such

37. Kripke, *Naming and Necessity*.
38. See ''Modalities and Intensional Languages'' and ''Quantification and Ontology,'' this volume.

sentences, but there is no presumption about objective reference of nonlogical terms. For such a substitutional account one starts not with reference but with truth. One might assign true to the sentence 'Pegasus is a winged horse' in a given world, and the existential generalization to 'Something is a winged horse' will mean nothing more than that some substitution of a syntactical item, for example, the syntactical name 'Pegasus' for '*x*' in '*x* is a winged horse' generates a sentence that is assigned true. On a substitutional possible-world semantics, initial truth assignments to atomic sentences replace domains and object assignments to names and "satisfaction conditions" for atomic sentences. Such a view has certain interesting uses in a semantics for fictional or mythological discourse and for discourse about putative possibilia freed of "commitment" to mythical or possible objects. But it misses a metaphysical point. Identity, which is a feature of *objects,* cannot be defined in such a semantics. Intersubstitutivity of *syntactical* items *salve veritate* does not generate objects, which must be *given* if identity is to hold. The intersubstituting of 'Father Christmas' and 'Santa Claus' even if truth values are preserved no more generates identical *objects* than does the intersubstituting of 'not' and 'not not not', which also holds *salve veritate*. Substitutional semantics may have some uses for nonobjectual discourse, but, as I now believe, only in conjunction with objectual quantification for the domain of actuals.[39]

In summary, modal discourse need not and should not admit possibilia despite the elegance of the generalization. Dispensing with possibilia is grounded not in the unavailability of criteria of identification but rather in the fact that identity is a relation for objects already given. Putative possibilia are not fleshed out with that complement of properties, relations, and a locus in the actual order—or, if material, in the physical order—that would enable them to count as objects at all.

39. The use of substitutional semantics for nonobjectual discourse is discussed at greater length in my presidential address to the American Philosophical Association, Western Division, "Dispensing with Possibilia," *Proceedings of the American Philosophical Association,* vol. 49 (1976): 38–51.

14. A Backward Look at Quine's Animadversions on Modalities

In April of 1988, Washington University in St. Louis sponsored a conference, "Perspectives on Quine," organized by Robert Barrett and Roger Gibson. It was a retrospective as well as prospective occasion. The conference papers, including the one here, were published in *Perspectives on Quine,* ed. R. Barrett and R. Gibson (Oxford and Cambridge, Mass.: Blackwell, 1990), pp. 230–243. ■

In this paper I will reflect on W. V. Quine's animadversions on modalities, the debates they provoked, and some of the outcomes. My reflections will be loosely historical and personally reminiscent.[1]

Quine traces his disaffection with modal logic to what he saw as the primary motivation of C. I. Lewis in devising modal propositional logic, a motivation that he attributes to a central confusion about use and mention. What Lewis was seen as doing was appending to the familiar propositional calculus, operators, axioms, and rules that were supposed to capture notions like validity, logical truth, consistency, and, most particularly, logical consequence. That project was perceived as undermining advances made by Alfred Tarski, Rudolf Carnap, and others when they took those semantical notions as metalogical and *about* the features of an interpreted formal calculus. In 1962 Quine wrote:

> Professor Marcus struck the right note when she represented me as suggesting that modern modal logic was conceived in sin: the sin of confusing use and mention. She rightly did not represent me as holding that modal logic *requires* confusion of use and mention. My point was a historical one, having to do with Russell's confusion of 'if-then' with 'implies'.
>
> Lewis founded modern [propositional] modal logic but Russell provoked him to it. For whereas there is much to be said for the material conditional as a version of 'if-then' there is nothing to be said for it as a version of 'implies'; and Russell called it implication thus apparently leaving no place for genuine deductive connections between sentences. Lewis moved to save the connections. But his way was not, as one could have wished, to sort out Russell's confusion of 'implies' with 'if-then'. Instead, preserving that confusion, he propounded a strict conditional and called *it* 'implication'.[2]

Quine ruefully concluded on that occasion that the use-mention confusion seemed also to be "a sustaining force, engendering an illusion of understanding."[3]

---

1. For a sharply honed examination of Quine on the more inclusive subject of opacity, a study of David Kaplan's paper in The Library of Living Philosophers volume on Quine will repay the strenuous effort it demands. L. Hahn and P. Schilpp, eds., *The Philosophy of W. V. Quine* (La Salle, Ill.: Open Court, 1986).

2. "Reply to Professor Marcus," *Synthese,* XIII (1961): 323. Reprinted in *The Ways of Paradox and Other Essays* (New York: Random House, 1966). I have inserted quotes around the last occurrence of 'implication'. Also inserted is '[propositional]'.

3. Ibid., p. 324.

But clearly the judgment is harsh. Carnap in 1934 made a similar point, but with important differences. He wrote:

> Russell's choice of the designation 'implication' for the sentential junction with the characteristic TFTT has turned out to be a very unfortunate one. The words 'to imply' in the English language mean the same as 'to contain' or 'to involve'. Whether the choice of the name was due to a confusion of implication with the consequence relation, I do not know; but, in any case, this nomenclature has been the cause of much confusion in the minds of many, and it is even possible that it is to blame for the fact that a number of people, though aware of the difference between implication and the consequence relation, still think that the symbol of implication *ought* really to *express* the consequence relation, and count it as a failure on the part of this symbol that it does not do so.[4]

Carnap[5] at the same time continues to use 'implication' for the material conditional and does not use it for 'consequence'. He goes on to introduce a stronger conditional, which he calls 'L-implication', into his sample object languages, a conditional that is supposed to have the feature that '*A* L-implies *B*' is valid just in case '*A* implies *B*' is valid. On a special notion of equipollence, '*A* L-implies *B*' is supposed to be equipollent to the metalinguistic '' '*B*' is a consequence of '*A*'.'' Where an equipollence holds, Carnap calls the object-language correlates ''quasi-syntactic.'' When he moves from syntax to semantics he does not abandon those quasi-syntactic object-language constants. He continues to urge that, for sentences with nonextensional or modal expressions, ''it will still be convenient to translate sentences of this kind not only into the meta-language but *in addition or instead* into syntactical [object-language] sentences with respect to a suitably constructed calculus.''[6] Indeed, in sorting logical from nonlogical signs of a formalized language, Carnap includes among the logical signs those for what he calls the ''logical modalities,'' among them a sign for Lewis's strict implication.[7] Also, in sorting out projects for future study, as he was wont to do, he especially singled out formalized accounts of languages with signs for what he calls ''logical, physical,

---

4. *The Logical Syntax of Language* (New York: Harcourt, Brace, 1937), p. 255. The earlier German edition was 1934. Italics mine.
   5. Ibid. See especially §§63–69; §69 concerns the logic of modalities.
   6. *Introduction to Semantics* (Cambridge: Harvard University Press, 1942), p. 249.
   7. Ibid., §13.

and causal modalities.''[8] Now, Carnap was no use-mention confuser. Still, he granted that there were good grounds that justified formal treatments of modalities. What remained was to arrive at a plausible semantical account of modal languages.

As we know, early intuitions about modal logic were too shaky even to ground a choice among some of the alternative Lewis systems, to say nothing of the added complexities of extending such systems to include, as I did,[9] quantification theory, identity, a theory of types, and no presumption of extensionality for properties. But the absence of an adequate semantics was not a deterrent. It is well to keep in mind that *Principia Mathematica,* authored by those regarded as use-mention confusers, long preceded formalized semantics.

Quine's scoldings diminished but did not stem the interest in modal logic. It continued, for one reason or another, to engage the attention of some logicians and philosophers, including exacting mathematical logicians.[10] Granted that, for a long time and under Quine's influence, it was considered by many philosophers to be misdirected and misconceived.

A footnote to Quine's historical point about historical origins: It is, I believe, inaccurate to claim that the preoccupation with the search for an object-language surrogate for deducibility was Lewis's only motivation. Consider the following quotation. The author is discussing Leibnizian possible worlds and says:

> This conception of possible worlds is not jejune: the actual world as far as anyone knows it, is merely one of many such which are possible. For example I do not know at the moment how much money I have in my pocket, let us say it is thirty cents. The world which is just like this one except that I should have thirty-five cents in my pocket is a consistently thinkable world. . . . An analytic proposition is one which would apply to or be true of every possible world. . . .[11]

That is C. I. Lewis in 1943, not David Lewis.

By the 1950s Quine begins to give a millimeter. He now allows

---

8. Ibid., p. 243.

9. ''A Functional Calculus of First Order Based on Strict Implication,'' *Journal of Symbolic Logic,* XI (1946); ''The Identity of Individuals in a Strict Functional Calculus of Second Order,'' ibid., XII (1947).

10. Among those who wrote on modal logic before 1960 were A. Bayart, R. Carnap, A. Church, H. B. Curry, M. Dummett, F. Fitch, K. Gödel, S. Kanger, E. J. Lemmon, J. Lukasiewicz, J. C. C. McKinsey, A. Prior, B. Sobocinski, A. Tarski, G. H. Von Wright.

11. ''The Modes of Meaning,'' *Philosophy and Phenomenological Research,* IV, 2 (1943–44): 236–249. Reprinted in *Collected Papers* (Stanford: Stanford University Press, 1970), pp. 303–316. Quotation is on p. 310.

that "there are three different degrees to which we may allow our logic or semantics to embrace the idea of necessity." [12] As it turns out, these embraces range from cool to stone cold. The first degree takes a modal term as a semantic predicate that attaches to sentence names, although it may not be exactly reducible to validity. The second, more "drastic" degree takes necessity as operating on closed sentences, suggesting *de dicto* use. The third, most "grave" degree occurs where, like the negation sign, it operates on open sentences, as in '□(x is a man)'. Although difficulties are noted for the first degree, it is the second and especially the third that in the late 1940s became the primary target and remain the target. However, over the years, some criticisms have been withdrawn or tempered, new criticisms have emerged, and finally the whole thrust shifts from a concern with motivation, paradoxes, puzzles, and the senselessness of interpretations of modal logic to claims about modal logic's repugnant essentialist commitments.

One can get a sense of the changing shape of the arguments as one moves from "The Problem of Interpreting Modal Logic" [13] in 1947 through the 1953, 1961, and 1980 versions of the paper "Reference and Modality." [14] It should be noted that Quine's more recent preoccupation has shifted away from modal quantification theories like those of Church [15] and Carnap [16], grounded as they were in a background Fregean structure that takes nonlogical terms such as singular terms to be systematically ambiguous; what these terms refer to and what they mean will alter with context. In the overworked example, 'The Evening Star' and 'The Morning Star' in

(1)   The Evening Star = The Morning Star

refer to one and the same thing, but in

(2)   John believes that The Evening Star = The Morning Star.

12. "Three Grades of Modal Involvement," in *Proceedings of the XIth International Congress of Philosophy 1953,* vol. 14 (Amsterdam: North-Holland, 1953). Reprinted in *The Ways of Paradox* (New York: Random House, 1966). Quotation is from the latter, p. 156. Quine, citing the interdefinability of the modal operators, confines the discussion to the necessity operator.

13. *Journal of Symbolic Logic,* XII, 2 (1947).

14. Included in *From a Logical Point of View* (Cambridge: Harvard University Press, 1953). Revised 1961 and again 1980.

15. "A Formulation of the Logic of Sense and Denotation," Abstract, *Journal of Symbolic Logic,* XI (1946): 31. Paper in P. Henle et al., eds., *Structure, Method and Meaning* (New York: Liberal Arts Press, 1951).

16. "Modalities and Quantification," *Journal of Symbolic Logic,* XI (1946); *Meaning and Necessity* (Chicago: University of Chicago Press, 1947).

those terms shift reference to nonidentical senses for Frege or to something like nonidentical senses or concepts or intensions for Church and Carnap. The swelling of ontology was for Quine a *prima facie* ground for rejection. For me, the ground for rejection was the systematic ambiguity.

I do not know whether Frege also thought that modal operators generated opaque contexts. Although there appear to be similarities with respect to *some* substitution failures as between modalized sentences and nonmodalized sentences with verbs for propositional attitudes, the presence of an agent as subject in the attitudinal assertions marks an important difference. Modal sentences or propositions such as

(3)   □(The Evening Star = The Evening Star)

(4)   □(The Evening Star = The Morning Star)

are not related to an agent's beliefs or attitudes. Although the move from (1) and (3) to (4) is also claimed to be a move from truth to falsity, as in analogous belief contexts, those differences will make a difference.

Quine suggests that the quotation contexts are the paradigm for opacity, as in

(5)   John says 'The Evening Star = The Morning Star'.

but here even a shift to intensions is precluded. The independent meaningfulness of any proper part of (5) in the scope of 'says' is erased by quotation.

There was a time for Carnap[17]—perhaps still for Davidson[18]— when belief sentences like (2) (John believes that the Evening Star is the Morning Star) were analyzed as affirming a relation between a subject and a sentence. It seemed oddly prejudicial that Quine continued to press the quotation analogy as the least objectionable approach to modalities (degree one involvement) but did not press it for propositional-attitude sentences, with which it has far greater affinity.[19]

---

17. *Meaning and Necessity*, §13.

18. Cf. "Thought and Talk," in *Mind and Language*, ed. S. Guttenplan (New York: Oxford University Press, 1975).

19. A. Church in "On Carnap's Analysis of Statements of Assertion and Belief," *Analysis*, X, 5 (1950): 97–99 (reprinted in *Reference and Modality*, ed. L. Linsky [New York: Oxford University Press, 1971]), pointed to some serious difficulties with such an account. That sentential account has resurfaced in another form, as in Fodor, where belief relates an agent to a sentence in "the language of thought."

The more substantive arguments Quine mustered over the years were, as noted, directed against second- and third-degree modal involvement. The criticisms are interwoven, but they can roughly be sorted into those on the one hand that might be seen as internal difficulties leading to puzzlement, bafflement, senselessness, and the like when one tries to interpret sentences with modal operators and quantifiers, and those on the other hand that maintain that even if claims of puzzlement were set aside or resolved, modal logic would still commit us to some undesirable and unacceptable philosophical or ontological or semantical views. Of the latter kind are, for example, a purported forced shift to intensions or a commitment to "essentialism." The background assumption of my modal quantification theory and of those that followed along the same lines was that there was no need to make a Fregean shift or to suppose systematic ambiguity. It was that non-Fregean context within which much of the debate fell. A forced Fregean shift relative to context would itself be a criticism and a sufficient deterrent. I took reference as univocal in and out of modal contexts.

Included in the second grade of modal involvement are sentences of modal propositional logic, as in the Lewis systems. Not much remains of Quine's criticisms of modal propositional logic beyond the dubious charge of misguided motivation. Occasionally Quine mentions, for example, unsettled views about acceptable postulates governing modal operators within the scope of other modal operators, as in '$\Box\Box A$' or '$\Box(A \supset \Diamond B)$' and the like, but the unsettledness is regarded as a symptom and is not explored independently.[20]

Still the latter is of interest, given some historical analogies between the extension of nonmodal sentential logic to include quantification theory and the extension of quantification theory to include modal operators. Close formal analogies hold between the universal and existential quantifiers and the operators for necessity and possibility.[21] There was in the early history of standard predicate logic also an occasional unsettledness about quantifiers in the scope of other quantifiers, especially when the variables of quantification varied, as in '$(x)(Ax \supset (\exists y)Fy)$'. It has been plausibly claimed[22] that Wittgenstein was baffled by such iterations. It was also sometimes argued that quantifiers are otiose, since they can be defined in terms of conjunction

---

20. "Three Grades of Modal Involvement," pp. 168–171 of *The Ways of Paradox*.

21. Indeed David Lewis takes those operators as quantifiers of a kind. See *Counterfactuals* (Cambridge: Harvard University Press, 1973).

22. Robert Fogelin, *Wittgenstein* (London: Routledge & Kegan Paul, 1987); see chap. 6, pp. 78–82. Fogelin's claims have been challenged, but they are plausible.

and disjunction or, taking classes as primitive, in terms of relations between classes. An adequate semantics for resolving those occasional perplexities about quantifiers in standard logic came well after *Principia Mathematica* in the work of Tarski and Carnap. Formal systems of quantified modal logic were similarly developed in the late 1940s and preceded semantical accounts. Stig Kanger's[23] semantical strategy for modal logic was not published until 1957 and unfortunately was not known by the likes of me. Saul Kripke's[24] and Jaakko Hintikka's[25] semantics for quantified alethic modal logic were not published until 1963. What the debate about iterated modalities might come to was articulated in those studies.

As noted above, the fact of conflicting intuitions about iterated modalities was not enlisted by Quine as a substantive argument against *propositional* modal logic. In fact, on the level of propositional calculus there remained no substantive criticism other than the specter of use-mention confusions. The serious quarrel was with quantified modal logic. In his autobiographical remarks[26] Quine speaks of my early papers[27] as taking up the "challenge," as did Carnap[28] and Church.[29] Since Carnap's and Church's background theory presupposed reference shifts to intensions and were unacceptable on that account, the primary target became modal quantification theory as given in my early papers. Quine graciously notes in his review of the first of those papers (1946) "the absence of use-mention confusions . . . a virtue rare in the modality literature."[30] But, alas, I fell from grace on the matter of use and mention in 1962, with a symposium paper to which Quine replied.[31]

23. *Provability in Logic* (Stockholm: Almquist and Wiksell, 1957).

24. "Semantical Considerations on Modal Logic," *Acta Philosophica Fennica* (1963). Reprinted elsewhere. A semantics for S5 appears in the *Journal of Symbolic Logic*, XXIV (1959): 1–15, 323–324.

25. "The Modes of Modality," *Acta Philosophica Fennica, Proceedings of a Colloquium in Modal and Many Valued Logics*, XVI (1963). Reprinted elsewhere. Hintikka had also proposed a semantics for quantified deontic logic in "Quantifiers in Deontic Logic," *Societas Scientarum Fennica*, IV (1957).

26. *The Philosophy of W. V. Quine*, p. 26.

27. See note 9. Those papers were reviewed by Quine in the *Journal of Symbolic Logic*, XI (1946): 96 and XII (1947): 95, with a correction of his 1947 review in XXIII (1958): 342.

28. See "Modalities and Quantification" and *Meaning and Necessity*.

29. See note 15 above.

30. See note 27 above.

31. That symposium was sponsored by the Boston Colloquium in the Philosophy of Science and took place in February of 1962. In conjunction with that symposium, "Modalities and Intensional Languages" (this volume) and Quine's comments were published in *Synthese* 1961 and in M. Wartofsky, ed., *Boston Studies in the Philosophy of Science 1961/1962* (Dordrecht: Reidel, 1963). Also included in this volume and in *Boston Studies*

Returning now to Quine's response to the challenge of quantified modal logic, I should like to touch on some features of my systems of quantified modal logic that were especially relevant to the debate. The modal propositional base was Lewis's S2 extended to S4 and S5. S4 and S5 include axioms for iterated modalities. The abstraction operator is seen as forming attributes, not sets. The latter, i.e., sets, are viewed as collections of just those things that have the attribute. The axioms are given as schemata. The nonmodal base is type-theoretic second-order quantification theory with individual constants, an abstraction operator, no description operator, and identity defined. A modal axiom is included about mixture of modal operators and quantifiers. Two metatheorems should be mentioned, since I believe that an appreciation of them would have deflected some of Quine's subsequent criticisms about the perplexities engendered by modal quantification theory.

The first theorem states that, for the material biconditional, substitution is restricted to nonmodal contexts. For the strict biconditional, substitution is unrestricted in modal contexts in both S4 and S5.

The second is the theorem about the necessity of identity:

NI $\quad (\alpha)(\beta)((\alpha = \beta) \supset \Box(\alpha = \beta))$

where 'α' and 'β' are syntactic notation for individual variables. NI also holds where '$\supset$' is replaced by '$\dashv$'.

A tangential remark: As I proceed with the review of Quine's arguments, there is a sense of *déjà vu,* since I will be going over some of the same ground covered in the above-mentioned 1962 symposium at Harvard where Quine commented. Present at the time were Saul Kripke, still an undergraduate, and Dagfinn Føllesdal, who had recently completed graduate study. Like the one whose namesake I am, I stood in alien corn, and I appreciated Kripke's support during the discussion of what he took to be some of my views. I began with some general considerations in justification of modal logic. In summary I urged that motivations went far beyond the search for an object-language correlate of validity and consequence. Modalities, including causal and physical modalities, are firmly entrenched in common and scientific language. I also took it as noncontroversial that our ordinary discourse, like scientific discourse, presupposes that there are things: fixed objects of reference. In formal semantics things are what constitute the domain of the interpretation of a language, as Quine used

is a much-edited "Discussion" culled from a tape of the original discussion, which lists Quine, Kripke, Føllesdal, McCarthy and me as participants.

to remind us. I urged more generally that "things" are what may properly enter into the identity relation.

If we are to talk about things in a public language, if we are to entertain the possibility that a thing might not have had the properties that in actual fact uniquely describe it, or if a thing remains what it is through many vicissitudes but ceases to exist altogether through others, then there is a semantical role for genuine proper names that is different from the semantical role of singular descriptions. I called this directly referring role "tagging" and said that beyond that role those names have no meaning. Singular descriptions can, as Russell explains in *Principia Mathematica,* be surrogate for genuine proper names in suitably restricted extensional contexts—where taking them as flanking an identity sign instead of unpacking them in accordance with the theory of descriptions will not, on substitution, take us from truths to falsehoods.

I pointed out that

> it often happens in a growing changing language, that a descriptive phrase comes to be used as a proper name—an identifying tag— and the descriptive meaning is lost or ignored. Sometimes we use certain devices, such as capitalization with or without dropping the definite article to indicate a change in use.[32]

Capitalization, like the artifice of single quotes, is commonly used to deny independent meaning to those parts of an expression that are ordinarily taken as contributing to the meaning of the whole.

Against those recollections consider Quine's argument about the Evening Star and the Morning Star, which is supposed to render modal logic incoherent. It is interesting to trace the arguments through "The Problem of Interpreting Modal Logic" in 1947 and the 1953, 1961, and 1980 versions of "Reference and Modality."[33] In the 1953 and 1961 versions Quine acknowledges Arthur Smullyan's[34] proposed solution, which purports to employ the theory of descriptions, but criticizes Smullyan's employment as "an alteration of Russell's logic of descriptions" because "Smullyan allows differences of scope to affect truth value even in cases where the description concerned succeeds

---

32. See pp. 10–11, this volume. The special semantic role of proper names is also discussed in my paper "Extensionality," *Mind*, n.s., LXIX (1960).

33. See notes 13 and 14 above. In the 1953 version of "Reference and Modality" see p. 155 and note 9. In the 1980 version, see p. 154 and note 9 in *From a Logical Point of View*.

34. "Modality and Description," *Journal of Symbolic Logic*, XIII (1948). Reprinted elsewhere. See my review, Appendix to "Modalities and Intensional Languages," this volume.

in naming.'' But of course it was a mistake to claim that Smullyan had ''altered'' Russell's logic of descriptions. It was in fact an exact employment of that logic as laid out in ''On Denoting'' and exemplified there by Russell in his analysis of apparent substitution failures in contexts of epistemological attitudes. It is a central point of Russell's theory that in such contexts even singular descriptions that succeed in denoting one thing must be unpacked. Quine's claim in 1953 and 1961 that for Russell ''change in the scope of a description was indifferent to the truth value of any statement unless the description failed to name'' was false and missed what Russell regarded as an innovative feature of the theory. In the 1980 revision those passages about Smullyan's purported misemployment of the theory of descriptions are deleted and replaced. Now Smullyan's solution is represented as ''taking a leaf from Russell,'' and it is seen that ''scope is indifferent to extensional contexts. But it can still matter in intensional ones.''

In summary, by 1980 Quine has finally agreed that by fully employing the theory of descriptions and allowing for ordinary proper names, the substitution failure is dispelled. But now he points out that the successful analysis places a modal operator in the scope of a quantifier and in front of an open sentence, which means ''adopting an invidious attitude toward certain ways of uniquely specifying [an object]x'' and ''favoring other ways . . . as somehow better revealing 'the essence' of the object.'' [35] So the issue is now the specter of essentialism: a sorting of properties as essential or nonessential to objects that have them.

Some further comments on the arguments from substitution failure and essentialism. As Quine actually presented it, 'The Evening Star' and 'The Morning Star' *are* capitalized. If the descriptions have thereby been converted into proper names, there is, in accordance with the necessity of identity (NI), no substitution failure. Substitutivity of proper names goes through in modal contexts *salve veritate*. So here Quine would have to question direct fixed reference for ''genuine proper names'' or the plausibility of necessity of identity or both.

One of Quine's early criticisms *was* directed against the necessity of identity (NI). In 1953 he claims that NI *forces* a shift from extensions to intensions. What went unnoticed is that one of the powers of the theory of descriptions is that it unpacks so-called contingent identities that generally require at least one singular description into

---

35. The 1980 version of ''Reference and Modality,'' in *From a Logical Point of View*. See note 14 above.

material equivalences that cannot be intersubstituted within the scope of the necessity operator. Indeed, the Smullyan solution to the substitution puzzle falls under the general substitution theorems for biconditionals mentioned above.

By 1961 the claim that modal contexts generally and NI in particular *force* a replacement of extensions by intensions has been abandoned. Instead Quine says that my preparedness "to accept . . . essentialist presuppositions seems rather hinted at."[36] by NI. The specter of essentialism again! By 1962 concerns about NI seem to vanish altogether so long as we stick to variables. Quine notices that the proof of NI requires only the necessity of self-identity and substitutivity for identity. What could be less controversial?

On the matter of proper names Quine sees more serious difficulties. He says:

Prof. Marcus developed a contrast between proper names and descriptions. . . . I see trouble anyway in the contrast . . . as Prof. Marcus draws it. Her paradigm of the assigning of proper names is tagging. We may tag the planet Venus some fine evening, with the proper name, 'Hesperus'. We may tag the same planet again some day before sunrise with the proper name 'Phosphorus'. When at last we discover that we have tagged the same planet twice our discovery is empirical. And not because the proper names were descriptions.[37]

Among considerations informing my view was the claim of linguists that proper names are not lexical items at all. They lack "lexical meaning."

Quine saw trouble. I did not. Empirical discoveries do not identities make. So, even on the matter of supposing as I did that there were directly referring proper names, it appeared that for Quine the trouble also came down to essentialism, since it suggested that things have their proper names necessarily. During the discussion that ensued after Quine's comments Kripke reinforced Quine's view with his remark that "such an assumption of names is equivalent to essentialism."[38] But that was not *my* claim. Socrates might have been *named*

36. *From a Logical Point of View*, the 1961 and 1980 revisions, p. 156. See note 14 above.

37. "Comments," *Boston Studies in the Philosophy of Science* (Dordrecht: Reidel, 1963), p. 101. (Reprinted in *Ways of Paradox* as "Reply to Professor Marcus," p. 179.)

38. R. B. Marcus, W. V. Quine, S. Kripke, J. McCarthy, and D. Føllesdal, "Discussion," p. 116 [appendix 1A, this volume, p. 35].

In response to the query about determining who or what was the reference of a proper name, I said one might look in a dictionary. I had in mind something like a biographical

Euthyphro; he would not thereby *be* Euthyphro. My claims about the special semantical role of proper names reemerged more widely in the early 1970s in theories of direct reference.

As we trace Quine's arguments over time, we notice that most of them are dispelled as originally proposed and are replaced by arguments about modal logic's commitments to essentialism. Since it does not require modal logic to note that, for example, being self-identical and being married to Xantippe are different kinds of properties that Socrates had, one expects a more *direct* attack on what seems on the face of it a very plausible sorting of properties.

There are two arguments in which Quine attacks such a sorting of properties directly. The first, in *Word and Object* (1960), has to do with the bewilderment that is supposed to be induced by the premises (1) Mathematicians are necessarily rational and not necessarily two-legged, and (2) Cyclists are necessarily two-legged and not necessarily rational, and (3) Someone, call him John, is both a cyclist and a mathematician. Quine asks, "Is this individual necessarily rational and contingently two-legged or vice versa?" and concludes that "there is no semblance of sense in rating some of his attributes as necessary and others as contingent."[39]

Of course the conditional premises here are ambiguous as to where the modal operator is located. In the first premise, are we saying '$\Box(x)(Mx \supset Rx)$' or '$(x)(Mx \supset \Box Rx)$' (we assume for simplicity the Barcan formula)? If the proper reading is the latter, with the modal operator attached to the consequent, the premises (1) (2) (3) are inconsistent. If, more plausibly, the proper reading is the former, then nothing baffling follows. That argument, presented in 1962, does not surface again in Quine's later critiques of modal logic.

A second direct assault seems to be that extensionality for sets is threatened by quantified modal logic, for shouldn't the set of creatures with a kidney and the set of creatures with a heart be identical, even though the equivalence between them is the set analogue of the material biconditional? The answer to this is yes, but the matter is also more complicated, and I have addressed it elsewhere.[40] In summary,

---

dictionary or those parts of a dictionary so headed, or like the *Oxford Classical Dictionary*, which provides information about the objects named.

The direct-reference theory for proper names is bolstered by the fact that linguists generally do not take proper names as lexical items at all. The absence of "content" makes them recalcitrant to "definition." [Added 1991.]

39. *Word and Object* (Cambridge: M.I.T. Press, 1960), p. 199. On my quantified modal logic, permutation holds for '$\Box$' and the universal quantifier, which is relevant to the summary of the argument.

40. "Classes and Attributes in Extended Modal Systems," *Acta Philosophica Fen-*

sets may be taken to be collections of objects satisfying various descriptions. Being uniquely a first star of the evening and being uniquely a first star of the morning are weakly equivalent attributes, but the objects that satisfy both are, if identical, necessarily so. Identity for sets (what I call "assortments") is an extrapolation of the identity of individuals, and, given NI, if such collections are identical, they are necessarily so. The attribute, as in the case of singular descriptions, may be seen, using Kripke's phase, as "fixing the reference." What finally seems to remain of Quine's critique is again essentialism, yet without any head-on account of what makes it invidious.

What, then, *is* invidious about what Quine calls "Aristotelian essentialism"? Aristotle has complex theories about universals, particulars, substances, species, essence, and accident. Furthermore it has been persuasively argued that Aristotle's modalities are temporalized, and relative to the actual.[41] Still there are features of his views that illuminate the debate and are related to what contemporary authors have called "metaphysical" modalities.

For Aristotle, a particular object that has independent existence is directly referred to as "a this." Particulars like Socrates or the Sun cannot be "defined" by specification of a unique set of Aristotelian essential properties. Even with respect to the Sun, which Aristotle took to be unchanging and uniquely and eternally rotating around the earth, he says, speculatively, that if it were to stop moving it would still be the Sun, and if another body were to rotate in the same path around the earth it would not be the Sun.[42] One can only suppose here that Aristotle also had some views of modalities that are not temporal and that are not made explicit.[43]

For Aristotle, essential properties are sortal. Objects which have them, have them necessarily, but there are objects that do not or might not have them. Being an entity is a nonsortal property, and so, presumably, is being self-identical. Also seemingly excluded are properties that are uniquely sortal, such as being identical to Socrates,

nica, Proceedings of a Colloquium in Modal and Many Valued Logics, XVI (1963), and "Classes, Collections, Assortments, and Individuals," this volume.

41. See S. Waterlow (Broadie), *Passage and Possibility: A Study of Aristotle's Modal Concepts* (New York: Oxford University Press, 1982).

42. *Metaphysics,* VII.15, trans. W. D. Ross, in *Collected Works of Aristotle* (New York: Random House, 1941).

43. I have been helped recently in threading my way through some of Aristotle's texts by J. Lear, *Aristotle: The Desire to Understand* (New York: Cambridge University Press, 1988).

but that are parasitic on nonsortal necessary properties such as self-identity.

There are in Aristotle additional constraints on essential properties of particulars, such as sorting things from their parts, but we need not consider those elaborations. Now, surely Quine can have no objection to such nonsortal necessary properties or to those parasitic on the latter, for they correspond to those predicates that are formed from logically valid sentences. A device for forming predicates from sentences does not exclude the valid sentences as sentences from which predicates can be formed.

Curiously, Quine's examples of repugnant essential properties are just such noncontroversial necessary properties as self-identity, and he characterizes essential properties in such a way as to include them.[44] In a recurrent example he uses, the property of being self-identical is taken as necessary and the property of being self-identical while *P,* where '*P*' is some contingent truth, is taken as contingent. Quine says that *A* is an essential property and *B* is a contingent property where

QE $\quad (\exists x)(\Box Ax \cdot Bx \cdot \sim \Box Bx)$

But that is not sufficient for troublesome essentialism [Take '$\Box Ax$' as '$\Box(x=x)$'.]

In a paper on essential properties[45] I mentioned candidates for characterizing Aristotelian essential properties. The crucial one of such candidates in the notation of my quantified modal logic says of an attribute $\hat{y}Ay$ that it is essential where

ES $\quad (\exists x)(\exists z)(\Box(\hat{y}Ay(x)) \cdot \sim \Box(\hat{y}Ay(z)))$

I argued that, on ES, modal logic accommodated talk about such properties but that there seemed to be no *commitment* to them in the sense that on interpretation an irreducible essentialist truth was a consequence of the axioms and rules.

Terence Parsons[46] extended ES to include *n*-adic predicates (excluding trivial parasitic ones) and proved that there were models of

---

44. "Three Grades of Modal Involvement," in *Ways of Paradox.* See especially p. 176.
45. "Essentialism in Modal Logic," *Noûs,* I (1967), and this volume. These arguments were presented more informally in my "Modalities and Intensional Languages" (1961), this volume. See note 31 above.
46. T. Parsons, "Essentialism and Quantified Modal Logic," *Philosophical Review,* LXXVIII, 1 (1969). An additional constraint excludes predicates like "identical to Socrates."

modal quantification theory consistent with taking as false all instances of ES extended to *n*-adic predicates. But there are also models of modal logic consistent with the truth of nontrivial essentialist claims. Models with truths of the essentialist kind, as in ES, go beyond the purely logical necessities, and this is perhaps the juncture at which one might say that the metaphysical modalities begin. Indeed, assumptions like the essentiality of being of a natural kind would seem to have to be imported.

I am not agreeing that metaphysical modalities are invidious. They have an important role in common as well as scientific discourse. A particular instance of a species such as tiger or of a physical kind such as gold would seem to have non-purely-logical necessary properties, as was argued by Aristotle, as well as by contemporary philosophers.[47]

It may be that what concerns Quine has finally little to do with interpretations of modal logic or use-mention confusions or sense-lessness or even essentialism. It may be that what lies behind his concern is that modal logic presents one of the challenges to his attack on *any* analytic-synthetic distinction where even logical truths are included among the analytic truths. Recall that, on Quine's characterization of essential truths, predicates formed from valid sentences, like being green or not green or being self-identical, are also to be included among the predicates he takes to be invidious. Indeed, such predicates are among his recurrent examples.

Even if Quine were correct in claiming that there are *no* sentences, not even those we call logically valid, that justify sorting propositions or properties into necessary and nonnecessary where the sorting remains fixed over the history of its use, that claim should not be confused with the fact that at any time slice in the history of the language there will be such a sorting. I believe that the whole debate about the analytic-synthetic distinction has been blurred by failure to distinguish between language viewed synchronically and language viewed diachronically (i.e., historically). The logical necessities remain so stable over time as to present the most obvious challenge to Quine's attack on the analytic-synthetic distinction, synchronically as well as diachronically.[48]

47. S. Kripke, "Identity and Necessity," in *Identity and Individuation*, ed. M. Munitz (New York: New York University Press, 1971), and "Naming and Necessity," in *Semantics of Natural Language*, ed. D. Davidson and G. Harmon (Dordrecht: Reidel, 1972). Both are reprinted elsewhere. Also H. Putnam, "The Meaning of 'Meaning' " in *Language, Mind, and Knowledge*, ed. Keith Gunderson (Minneapolis: University of Minnesota Press, 1975); reprinted elsewhere. See also my "Essential Attribution," this volume. The literature on these topics is now large. The "metaphysical" modalities are in some respects like Carnap's "physical" modalities.

48. Quine has not been wholly consistent in his claim that even the status of "logical

I have not discussed another notion related to essentialism, that of *individual* essences, and I do not propose to go on about them here, except to remind the reader of Aristotle's claim that particulars "cannot be defined." If we think of an essence not as an intensional object but as a finite set of nonindexical[49] general properties that uniquely selects a particular object from all others in all possible circumstances or, more metaphysically, across worlds, then there are strong arguments questioning the plausibility of such transworld identification. Kripke and others have sought such properties in *origins;* I doubt that such a search can succeed. Aristotle seems to me correct when he says that the "essence" (in the sense of essential properties) of a particular comprises those properties that it shares with all of that species and does not serve to distinguish further among particulars of the species.

In conclusion, a few tangential remarks. The debate has on the whole been productive. Quine's criticisms, while discouraging to some, had a way of stimulating interest in modal logic, modal semantics, theories of kinds, and physical modalities. Theories of direct reference finally flourished in the 1970s. Many who had once shared Quine's views adopted new ones as the force of his arguments diminished. The work of Dagfinn Føllesdal is an interesting example. In 1966 Føllesdal made available a monograph that was a revised version of his 1961 Harvard dissertation.[50] There he suggested three approaches to substitutivity failures of singular terms in modal contexts. The first, the Quinean solution, was to convert all singular terms to predicates, as in 'Socratize'. The second placed a restriction on *which* singular terms may be intersubstituted in modal contexts. The third was to use Russellian description theory for all singular terms. The second solution seems to accommodate what Kripke later (1971) called "rigid designators," which include proper names as well as some descriptions. But it does not distinguish the special semantical role of proper names as compared with rigid descriptions. Føllesdal discusses the advantages and disadvantages of each approach. The first is cumbersome. The second requires *inter alia* that what counts as a

---

truths" is "open to revision." In *Philosophy of Logic* (Englewood Cliffs, N.J.: Prentice-Hall, 1970; revised 1986) he argues that to change the meaning of the logical constants is to "change the subject."

49. I exclude predicates formed from sentences with indexical terms like 'here' and 'I' where indices remain in the predicate. Proper names are taken as quasi-indexical; hence, as in my paper "Essentialism in Modal Logic," being identical to Socrates is a quasi-indexical not a *general* property. In that paper I called such quasi-indexical properties "referential properties" in that they have components that refer to individuals directly. See note 45 above.

50. *Referential Opacity and Modal Logic* (1966), privately distributed. See especially the preface and §§19 and 20. The quotes are on pp. 102 and 120.

singular name vary as we move from extensional to modal contexts. The third, Russellian descriptions, is seen to have as its "main drawback" problems of substitutivity that, Føllesdal says, "could be avoided if we base our theory of descriptions on a set of contextual definitions different from that of Russell." Had a special preformal role for proper names of direct reference been recognized at that time, some of the difficulties with the second solution would have been dispelled. As for the third solution, Føllesdal was, like Quine, mistaken in the claim that Russell's theory would have to be revised. Føllesdal's preference is for the first Quinean solution: Socratize. But what Føllesdal does not reject in 1966 is the importance of preserving modal distinctions, and, in this respect, he has gone beyond Quine. He bends his efforts toward "making sense" of "Aristotelian essentialism" (Quine's version) and says that "to make sense of Aristotelian essentialism and to make sense of open sentences with an 'N' prefixed are one and the same problem."

Føllesdal[51] in his 1986 paper for the Quine volume of the Library of Living Philosophers now endorses the special role of proper names and a theory of direct reference. There is no longer a problem of "making sense" of expressions with modal operators attached to open sentences or modal operators in the scope of quantifiers. He points to results such as mine and those of Terence Parsons that are about the absence of commitment to nontrivial necessary properties in some models of modal logic and, hence, the absence of a *commitment* to essentialism (of the non-Quinean sortal version), without at the same time denying the "meaningfulness" of essentialism. We have come a long way.

In all of the debates Quine has been a gadfly; in that he is in good company.

---

51. "Essentialism and Reference," in *The Philosophy of W. V. Quine*, ed. L. Hahn and P. Schlipp (La Salle, Ill.: Open Court, 1986).

15. *Some Revisionary Proposals about Belief and Believing*

In May of 1986 I presented a version of this essay as a lecture to the Collège de France. The lecture was revised for publication in *Philosophy and Phenomenological Research*, L (Supplement, 1990): 133–153. It was also included in *Causality, Method and Modality*, ed. Gordon Brittan (Dordrecht: Kluwer, 1991). This printing includes some corrections and nonsubstantive clarifications.

Since the essay was freestanding, familiarity with the earlier "Rationality and Believing the Impossible," included in this volume, could not be assumed, and there is significant overlap. But the present essay also departs from the earlier paper in substantial ways. Here the claim that we cannot believe impossibilities is urged as a conceptual revision and extension rather than as an elucidation of the extant concept of belief and the relation of believing. ■

There is consensus about some general conditions on a theory of believing and belief, such as (1) believing is a relation between a subject, the believer, and an object or set of objects as given in the grammatical form of the sentence, '*x* believes that *S*'; (2) beliefs, whatever they are, can be acquired, replaced, or abandoned; (3) beliefs enter, along with desires, needs, wants, and other particular circumstances, into the explanation of action; and (4) for some circumstances and for some beliefs, it is appropriate to describe a subject's beliefs as justified or unjustified, rational or irrational, and the like.

But such general features are in contrast to a tangle of unreconciled views that appear when one tries to flesh out a theory or give the concepts more content. There is disagreement about the nature of the belief state of the subject, the nature of the object of the believing relation, the efficacy or causal role of belief in shaping actions, and the role of language in an account of belief. There is disagreement about whether there can be unconscious beliefs, about where one draws the line between believing and acting, and about whether non-language-users can have beliefs. The inventory is large.

This tangle has led some philosophers[1] to claim that discourse about belief is folk psychology to be replaced by proper science. The language of belief, they say, will fall into disuse, just as the theory of humors as an account of emotions fell into disuse.

I should like in this paper to sketch an account that may resolve some of the disagreements. In so doing I depart from some received views,[2] and hence the account may seem revisionary as an explication of belief. But it is not so revisionary as to be wholly without precedent or flagrantly out of accord with features of our ordinary understanding.

What is central in the account here presented is the departure from the dominant, language-oriented accounts of belief, which take it that the objects of believing are always linguistic or quasi-linguistic entities such as Frege's propositions or "thoughts" or Davidson's interpreted sentences. Since Frege, the preoccupation with belief claims, belief

1. See S. P. Stich, *From Folk Psychology to Cognitive Science: The Case against Belief* (Cambridge: M.I.T. Press, 1983). Also, D. Dennett, "Beyond Belief," in *Thought and Object,* ed. A. Woodfield (Oxford: Oxford University Press, 1982).

2. The present paper in later sections amplifies and revises my "Proposed Solution to a Puzzle about Belief," *Midwest Studies in Philosophy: Foundation of Analytic Philosophy,* vol. VI, ed. P. French et al. (1981), pp. 501–510, and "Rationality and Believing the Impossible," this volume. Those papers focused on S. Kripke's "Puzzle about Belief" in *Meaning and Use,* ed. A. Margalit (Dordrecht: Reidel, 1979). The disquotation principle discussed below was set out by Kripke.

reports, which *are* linguistic, and the efforts of formal semanticists to provide a semantics for sentences with epistemological verbs have sometimes obscured our understanding.

We are concerned to give an account of *"x believes that S."* I should like to begin with a critical examination of language-oriented views.

Language-centered Theories of Belief and Some Difficulties with Such Theories

In "Thought and Talk," Donald Davidson[3] argues that '*x* believes that *S*' is equivalent to '*x* holds a certain sentence true' in a shared interpreted language. That sentence is '*S*' or some translation of '*S*' in the shared interpreted language. Believing is a conscious relation of subjects to their utterances. Davidson goes further and claims that even desires relate a subject to utterances. No language, then no desires or beliefs. And finally, a non-language-user, he says, cannot even have thoughts.

There is a further-stated baffling claim: that we cannot have thoughts, beliefs, and even desires, without the *concepts* of thought, of belief, and of desire. With respect to belief, Davidson says, "Can a creature have a belief if it does not have the concept of a belief? It seems to me it cannot and for this reason. Someone cannot have a belief unless he understands the possibility of being mistaken and this requires grasping the contrast between truth and error—true belief and false belief. But this contrast I have argued can emerge only in the context of interpretation [of a language]."

The view on reflection is implausible. Consider the following example: A subject, call him "Jean," and his dog, call him "Fido," are stranded in a desert. Both are behaving as one does when one needs and desires a drink. What appears to be water emerges into view. It is a mirage for which there is a physical explanation: such a mirage occurs when lower air strata are at a very different temperature from higher strata, so that the sky is seen as if by reflection, creating the optical illusion of a body of water in the distance. Both hurry toward it.

On what possible ground can we deny Fido a desire to drink, a

3. In S. Guttenplan, ed., *Mind and Language* (Oxford: Oxford University Press, 1975), pp. 7–23. Quotation is from pp. 22–23.

belief that there is something potable there? Jean and Fido *are* both mistaken, but only a language user, Jean for example, has the concept of a mistake and can report it *as* a mistake. That does not require on the dog's part a *concept* of belief, a *concept* of desire, a *concept* of truth and error. The preverbal child hears familiar footsteps and believes a person known to her is approaching, a person who perhaps elicits behavior anticipatory of pleasure. It may not be the anticipated person, and when the child sees this, her behavior will mark the mistake. But must there be some linguistic *obbligato* in the child if we are to attribute to her a mistaken belief or a disappointment? Must an agent have the *concept* of a mistake to *be* mistaken?

The important kernel of truth in such a linguistic view is that arriving at a *precise* verbal description of another's beliefs and desires is difficult, and especially so when the attribution cannot be verbally confirmed by the subject. We will not go far wrong in attributing thirst to the dog Fido, or the belief that there is the appearance of something potable. Whether we can attribute to the dog the recognition of *water* would depend in part on whether dogs can select water from other liquids to roughly the phenomenological extent that we can. That is an empirical question. Of course, to *attribute* a belief, a desire, a thought to oneself or others, or to assert that someone has a desire or a belief or a thought, requires language. But, in the example given, Fido and Jean need not be making verbal claims, vocalized or nonvocalized, about what they desire, what their thoughts and beliefs are. What this language-based view entails is that without a verbal *obbligato* or without an identifiable *linguistic* representation there are no thoughts, desires, or beliefs. Nor does the problem of correct belief attribution disappear among language users, particularly if linguistic confirmation from the believer is unavailable. It is, of course, considerably reduced.

To decline to attribute desires and beliefs to non-language-users is reminiscent of Descartes's declining to attribute pain to higher nonhuman animals despite the similarity with the causes of pain and with pain behavior in nonhuman animals and language users. The case of belief is analogous. Descartes argues that, in the absence of conscious introspective thoughts about our states, such as pain thoughts or belief thoughts, we do not have those pains and beliefs. In the revamped current version it is the absence of language, rather than of mind, that deprives a subject of thoughts and beliefs.

Not all who have language-oriented theories of belief take so strong a stand. Some, like F. P. Ramsey, saw the distinction, but attributed it to an ambiguity in the notion of belief. It is of interest to note that Ramsey[4] allows a sense of 'belief' in which we may, using his curious example, attribute a belief to a chicken who has acquired an aversion to eating a species of caterpillar on account of prior unpleasant experiences. Here again an overly rich attribution is difficult to avoid. We surely cannot attribute to the chicken the belief that the caterpillar is *poisonous,* but surely we will not go too far afield if we attribute the belief that the caterpillar is not for eating. And, indeed, if presented with a caterpillar that had the appearance of the despised kind but was in fact of an edible kind, the chicken would be mistaken about its inedibility.

Still, although acknowledging a nonlinguistic use of 'belief', Ramsey finally concludes that believing as it occurs in language users is so disparate from that of non-language-users that the term 'belief' is ambiguous. He goes on to say, "Without wishing to depreciate the importance of this kind of belief, . . . I prefer to deal with those beliefs which are expressed in words . . . *consciously* asserted or denied. . . . The mental factors of such a belief I take to be words spoken aloud or to oneself or merely imagined, connected together and accompanied by a feeling or feelings of belief. . . ." For Ramsey then, 'assenting to a sentence "*S*" ', 'asserting that *S*' and 'believing that *S*' are equivalent alternative usages. Ramsey takes those utterances, spoken aloud or to oneself, as *mental* factors; hence, the objects of belief, the *S* in '*x* believed that *S*', are events of a linguistic character, sentences spoken or thought. Those are the sentences toward which we have an assenting attitude, a feeling, for Ramsey. This attitude or feeling performs in Ramsey's work the role of 'holding true' in Davidson's.

The identification of the objects of believing with sentence-like objects has some familiar consequences. The believer, the subject, has those beliefs. But how does the subject have them? Ramsey singles out those "mental factors," sentences spoken aloud or to oneself or imagined. To say instead with Frege that they are the quasi-linguistic entities, the propositional *contents* of sentences, does not alter the picture in a helpful way. Such propositions mimic the structure of sentences. They have properties that interpreted sentences have, like truth and validity. Sets of them can be consistent or inconsistent. They can be contradictory. They can enter into the consequence relation, and so on.

---

4. *The Foundations of Mathematics* (New York: Humanities Press, 1950), p. 144. Ramsey also claimed that such introspective feelings are an insufficient guide when it comes to judging the *difference* between believing more firmly and believing less firmly.

Frege[5] had the view that propositions are the abstract nonmental contents of sentences toward which we have mental attitudes. But the mind-centered locus of the objects of belief is not wholly evaded. We "have them in mind" when we entertain them, believe them, disbelieve them. He did after all call them "thoughts." A recent account that claims to demystify Frege's propositions and is more explicitly language-centered may be described as the *computational model*.[6] Highly simplified, the subject is seen as having an internal register of basic concepts and basic sentences that are mental representations of actual sentences in his language. Syntactical rules for mental-representation sentences generate complex sentences; deductive rules generate consequences of sets of sentences. Mental-representation sentences are associated with mental analogues of yes-or-no responses on appropriate cues that will generate mental yes-or-no responses to more complex sentences on appropriate cues. The subject is said to believe that S just in case his correlated mental-representation sentence elicits a mental yes response. Jerry Fodor[7] presents such a view, which I have much simplified. He says straight out that attitudes toward propositions are in fact attitudes toward formulas in "mentalese," the language of thought, formulas that are internally codified and are correlated with the external sentences of a given language. Propositions have given way to sentences in mentalese. The objects of belief are linguistic entities placed squarely in the mind.

There are failings in such language-centered, wholly mind-centered accounts of belief. These accounts exclude belief attributions to non-language-users or, alternatively, insist that *if* one can make such attributions and if the manifestations of a public language are absent, the language of thought sententially organized must indeed be there in the nonverbal child or the dumb animal's mind or brain. Such accounts also create difficulties for making sense of unconscious beliefs. The exclusion of unconscious beliefs is explicit in Ramsey and, it would seem, in the Davidson of "Thought and Talk." For Fodor, there remains the question of what would count evidentially in the attribution of an unconscious belief to an agent. Is it unconscious assent to a sentence in mentalese?

5. See "On Sense and Nomination" and other essays in *G. Frege, Translations from the Philosophical Writings,* trans. P. Geach and M. Black (Oxford: Blackwell, 1952). The present discussion of propositions as linguistic entities mapped by sentences that "express" them does not apply to those more recent accounts of propositions as functions from worlds to truth values.

6. Computer scientists concerned with such "artificial intelligence" models actually use the language of belief in discussing their programs.

7. See *The Language of Thought* (New York: Crowell, 1975) and *Representations* (Cambridge: M.I.T. Press, 1981).

Language-centered views also tend to define rationality in terms of attitudes toward sentences (or propositions) that are consistent, contradictory, logically true, or related by deducibility and the like. Since it is agreed that rational, language-using agents as ordinarily viewed are not omniscient or perfect logicians, they may still be rational to a point yet fail to believe *all* the consequences of their beliefs, and hence may even come to hold true or assent to a sentence that is equivalent to a blatant contradiction. Where to draw the line and yet preserve the attribution of normal rationality is difficult to decide. But this is not an insurmountable problem. Nor am I suggesting that considerations of consistency and validity of inference are irrelevant to an account of rationality. The point is rather that there is a broader notion of rationality and irrationality that language-centered theories are incapable of accommodating. There is, for example, the irrationality of the subject who sincerely avows that *S,* or holds '*S*' true, but whose nonverbal actions belie it. Such cases need not be centered, although they often are, around questions of *akrasia*, such as that of the subject who sincerely avows that smoking is harmful yet continues to smoke. There are plausible psychological claims that our explicit avowals of belief, our sincerely stated claims about our own states, such as our desires and fears, or about the objects of our affections and disaffections, often do not serve us as beliefs are supposed to in the explanation of action. Actions may belie our most sincerely reported "beliefs." These are not cases of *deliberate* deception or insincerity and may have some explanation in theories of self-deception or false consciousness. But those latter theories are often grounded in the absence of an agent's conscious formulations in language of the contrary implicit beliefs that explain the dissonant actions.

Consider the subject who assents to all the true sentences of arithmetic with which he is presented and rejects the false ones; who can perform the symbolic operations that take him from true sentences of arithmetic to true sentences of arithmetic, and who also has toward them the belief feeling. Yet if you ask him to bring you two oranges and three apples, he brings you three oranges and five apples. He never makes correct change. Are his assents and assertions sufficient for ascribing to him correct arithmetic beliefs? Shouldn't nonverbal behavior also count as an indicator or a counterindicator of belief?

And then there is the obvious fact, alluded to in the example of the thirsty desert wanderers, that we often, very likely more often than not, do not consciously entertain propositions or sentences we hold true when acting, even when our actions are explicable as consequences of beliefs and desires. Language users may assert such "propositions" if they are asked why they are acting as they are. Indeed, being asked why we are acting as we are may lead us to discover or describe a belief that had never been verbalized. I usually walk a route to my office that is not the shortest and am asked why. It requires some thought. It isn't out of habit, I decide. I finally realize that I believe it to be the most scenic route. Verbalization as a necessary *condition* of believing precludes our discovering and then reporting what we may already believe.

An Object-Centered Account

We are concerned here with beliefs purported to be about the actual world, not about fiction or myth or the like. This is not to deny the use of "belief" locutions in discourse about fiction. Their role will be understood from the context. If I am asked what it was that was converted into a chariot by Cinderella's godmother, I might respond that "I believe it was a pumpkin that was converted into a chariot," but contextual cues make it clear that I am not making a historical claim about the actual world.

What follows is a sketch of the world-centered, object-centered account that may be better fitted to our understanding of some epistemological attitudes.[8] Believing is understood to be a relation between a subject or agent and a state of affairs that is not necessarily actual but that has actual objects as constituents. We may think of states of affairs as structures of actual objects: individuals as well as properties and relations. The structure into which those objects enter need not be an actual world structure. The state of affairs described by the sentence 'Socrates is human' is a structure containing Socrates and the property of being human. Since believing is taken to be a relation of an agent to a state of affairs not necessarily actual, the believing subject may also be related to the constituents of the structure.[9] Analogously, my ancestors may be structured as a set. I as a descendant am related to that structure and also to each of its constituents.

Believing has often been called a *propositional* attitude. On the present account, if we wish to retain the locution 'proposition' for an object of believing, that usage is atypical. Since a proposition is more

8. Bertrand Russell maintained throughout his work an object-oriented view of epistemological attitudes that is sometimes obscured by his use of the term 'proposition', which normally has a linguistic connotation. "Propositions" for Russell contain nonlinguistic constituents.

9. For Russell, believing relates the agent to the *constituents* of the proposition, not to the proposition. This suggests that one relation precludes the other, but it need not.

commonly viewed as a linguistic or quasi-linguistic entity, it is best to deploy other terms such as 'state of affairs' or 'structure'. One recognizes here a Russellian thrust.[10] In one of Russell's early accounts of epistemological attitudes, constituents of propositions are actual objects, including abstract objects such as properties and relations. One of the departures from Russell herein is that no reductionism for constituents need be supposed. Ordinary individuals, properties, and relations may be constituents of states of affairs.

On the subject or agent side of the relation we give a dispositional account.[11]

D: *x* believes that *S* just in case, under certain *agent-centered circumstances* including *x*'s desires and needs as well as *external circumstances*, *x* is disposed to act as if *S*, that actual or nonactual state of affairs, obtains.

Note the absence of a truth predicate in D. Actual or nonactual states of affairs are not truth bearers. If we employed the truth predicate as in "disposed to act as if '*S*' were true," *S would* have to be a linguistic or quasi-linguistic entity. It was Russell's continued use of truth and falsity as properties of his propositions that made the intrusion of Mt. Blanc into one of Russell's "propositions" so baffling to Frege.[12]

Ways in Which Such an Account Accommodates Some Natural Views of Belief

(1) On the proposed view speech acts, public or private, are only a part of the range of behavior that manifests belief; they are not, as they are in language-centered views, necessary conditions. Accord-

10. See the papers mentioned in note 2 above. Also, R. Chisholm, "Events and Propositions," *Noûs*, IV, 1 (1970): 15–24; the recent work of J. Perry and J. Barwise on "Situation Semantics," e.g., *Situations and Attitudes* (Cambridge: M.I.T. Press, 1983); N. Salmon, *Frege's Puzzle* (Cambridge: M.I.T. Press, 1986).

11. R. B. Braithwaite, in "The Nature of Believing," *Proceedings of the Aristotelian Society*, XXXIII (1932–1933): 129–146, has a dispositional account, but it is also a language-bound account. For Braithwaite *"x believes S"* is analyzed as follows: *S* (a proposition) must be *entertained*, and, under relevant internal and external circumstances, *x* is disposed to act as if *S* were *true*. Such a language-oriented dispositional account excludes unconscious beliefs, excludes beliefs of non-language-users, and supposes that believing always entails entertaining linguistic or quasi-linguistic objects. The dispositional account I am proposing also differs from Quine's in that Quine does not take states of affairs as objects of believing.

12. G. Frege, *Philosophical and Mathematical Correspondence* (Chicago: University of Chicago Press, 1980), p. 169.

ingly, beliefs can be attributed to non-language-users. Naturally non-language-users will fail to have beliefs that are possible only to language users—beliefs about language, for example. Linguistic items, whether type or token, are objects and can be constituents of states of affairs, but they are inaccessible to non-language-users *as* linguistic items. Non-language-users will, therefore, not have beliefs *about* describing or referring, about truth or falsity, validity or logical consequences, about grammar, and the like. Inference as a *psychological* phenomenon will be severely limited in non-language-users, since complicated inference would seem to require stating, describing, or reporting what we believe; setting it out in language. Prediction, deception, counterfactual speculation, and long-range planning of a certain level of complexity might also seem to require language, as would second-order beliefs (although there are recent empirical studies that claim that non-language-users can and do plan and make long-range decisions and practice deception). But that is not to deny beliefs to non-language-users altogether.

(2) There is ample evidence that when we act out of belief we need not precede or accompany such actions with verbal, sentential accompaniments. We need not be *entertaining before the mind* sentences or meanings. Jean, the desert wanderer, is racing to the water he believes to be out there. Both Jean and Fido have a belief in that they are related to a (nonactual) state of affairs and, given their circumstances, act accordingly.

(3) Such an account of belief views believing as a relation between a subject and a state of affairs not necessarily actual, where the subject, in the grip of psychological states such as wants and needs and in the presence of other circumstances will act as if that state of affairs obtained. Speech acts are *among* the acts that may and often do *manifest* a belief, and one such speech act when x believes S is that x may sincerely assent to a sentence descriptive of the state of affairs S. The account does not suppose that the act of sincere assent, even where it is evoked, *must* be an overriding indicator of belief.

A range of circumstances will evoke such a speech act of assent. The subject x may want to report his beliefs, to communicate them in language to others, to examine carefully what follows from them, to testify, and so on. If circumstances and desires are such that x wants to deceive, he may perhaps not assent to a sentence that describes his beliefs, even though that behavior would be counted as insincere. But deliberately denying a sentence that describes *what* one believes is not the only way a speech act may mislead others about one's beliefs. Given that speech acts are important in *reporting* beliefs, the agent's particular way of reporting is a function of local circumstances as well as of other beliefs that may not be shared by others. It is also a function of a subject's mastery of the language. A language user of minimal competence may not even be able to perform a speech act that describes the state of affairs to which he is in the believing relation, but he is in that relation nevertheless. Indeed, despite the widespread assumption of privileged access to one's own beliefs, it could (and does) happen that someone other than the agent is better able to report an agent's beliefs than the believer. Also, others can often assist a believer to describe more accurately the state of affairs to which he is in the believing relation; that is a common phenomenon of language acquisition.

(4) The proposed view accommodates the possibility of unconscious beliefs. These may be the very beliefs I have but do not or (on some psychological theories) cannot report in a speech act. Reporting brings them into consciousness.

(5) It is a feature of the proposed non-language-centered view of belief that it permits a more adequate and natural account of rationality. Language-centered theories tend to define rationality in terms of sentences or sets of sentences or their quasi-linguistic "contents." On a language-centered account a rational agent is one who, for example, will not assent to surface contradictions; for a perfectly logical agent, belief is closed under logical consequence (*pace* Dretske and Nozick).[13] Given the empirical fact that we are not faultless logicians, belief for a rational agent is closed under logical consequence to some acceptable level of complexity of proof. A *norm* of consistency for sentences we assent to is preserved. We abandon assent to sentences *known* to be inconsistent or necessarily false. Extended to inductive reasoning, such an account of rationality still focuses on a relation between accepted sentences and probable conclusions. But this language-centered account is an impoverished view of rationality. It lacks explanatory force. *Why* should we dissent from known contradictions or inconsistent sets of sentences? A computer would pay no price for "assenting," nor presumably would a brain in a vat.[14]

---

13. F. I. Dretske, "Epistemic Operators," *Journal of Philosophy*, LXVII (December 24, 1970): 1007–1023, and later R. Nozick, *Philosophical Explanations* (Cambridge: Harvard University Press, 1981), question the claim that belief is closed under logical consequence, but their purported counterexamples are not critical in our present account.

14. Andrew Hodges, in *Alan Turing: The Enigma* (New York: Simon and Schuster, 1983), p. 154, reports a conversation between Turing and Wittgenstein on contradiction that includes the following exchange:

WITTGENSTEIN (citing the paradox of the liar): It doesn't matter . . . it is just a useless language game. . . .

There is a wider notion of rationality than those of strongly language-centered views, where we say of a rational agent that such an agent also aims at making all the behavioral indicators of belief "coherent" with one another. For example, the agent's assentings and avowals should be coherent with his choices, his bets, and the whole range of additional behavioral indicators of belief. We may say of someone who avows that he loves another yet often harms the one he claims to love that his behavior is "dissonant." He is, in a wider sense, irrational. He is not logically irrational in the narrow, language-centered sense. The set of sentences he assents to are, so far as he knows, consistent. When he assents sincerely to a sentence 'I love *A*', he does not assent to a sentence 'I don't love *A*'. He would deny the latter if asked.

Agents who become aware of their incoherent behavior may try to "rationalize" that behavior, make it coherent, get a better fit. On such awareness the ambivalent lover may no longer assent sincerely to 'I love *A*'. He may note that other actions are incoherent with his speech acts, or he may alter his cruel behavior to fit his avowals. He may argue that the concept of love is confused. That list does not exhaust the possibilities.

Such considerations are usually viewed as central to questions of *akrasia*. An agent might sincerely avow that smoking is harmful and that he wants to preserve his health but continue to smoke. Recurrent in the literature on *akrasia* are explanations of such seemingly irrational behavior. One explanation is that the akratic has conflicting beliefs, one belief conscious and reportable in a speech act, one unconscious and unreported, which is the action-guiding belief that overrides if he continues to smoke. A second[15] explanation is that the akratic agent has conflicting reportable beliefs, and, although the acknowledged grounds that justify one are stronger than the grounds that

---

TURING: What puzzles one is that one usually uses a contradiction as a criterion for having done something wrong. But in this case one cannot find anything done wrong.

WITTGENSTEIN: Yes—and more: nothing has been done wrong . . . where will the harm come?

TURING: The real harm will not come in unless there is an application in which a bridge may fall down or something of that sort.

WITTGENSTEIN: . . . But nothing need go wrong, and if something does go wrong— if the bridge breaks down—then your mistake was of the kind of using a wrong natural law. . . .

TURING: Although you do not know that the bridge will fall down if there are no contradictions, yet it is almost certain that if there are contradictions it will go wrong somewhere.

15. D. Davidson, "How Is Weakness of the Will Possible?" in *Moral Concepts,* ed. J. Feinberg (Oxford: Oxford University Press, 1970).

justify the other, he acts in accordance with the less justified belief. A third sees akratic action as action not grounded in belief at all, but as "compulsive." Still, in all such explanations of irrationality it is our nonverbal acts that belie our words.

Nor are we suggesting that adding coherence to strict logicality is exhaustive of a wider account of rationality. There are psychological syndromes, paranoia for example, in which an agent's actions may be described as remarkably coherent albeit irrational. A still wider account of rationality must also include acting in accordance with norms of evidence, norms of justification, and the like that warrant an agent's believing as he does. Our proposal is only that coherence is an important feature of a more general explication of rationality.

Assentings and avowals do play an important role in a wider notion of rationality, for an agent is often the best describer of what he believes, and an external judgment of coherence or incoherence is more determinate, given such speech behavior. But, more important, a wider notion that includes coherence is explanatory of why a norm of logical consistency is preserved. Why, for example, is it claimed that a rational agent does not assent to a *known* contradiction?

If a sentence '*S*' describes a specific state of affairs and if sincere assenting to '*S*' is taken as an act, a speech act, which marks our believing that *S* obtains, then if, without relinquishing our original assent, we also assented to 'not-*S*' under the same circumstances, we would be believing that *S* both obtained and did not obtain in those circumstances—an objective impossibility that might render many of our actions incoherent and self-defeating.[16] Rationality in the wider sense is not preserved. If the assents to contradictions are not ferreted out, beliefs could lose their crucial role in guiding and explaining actions.

Belief, Assent, and the Disquotation Principle

The object-centered position sketched here—and it is just a sketch—does not preclude special relationships between some speech acts, such as acts of assent, and belief, under some conditions. What has been

16. It is difficult to make a case for rejecting contradictions unless we see the connections between rationality, coherent action, and plausible outcomes. A computer programmed with proper deductive rules and a contradiction will allow any sentence in its register of affirmations. In the absence of further action to be guided by those outputs, there is no problem of coherence as here described. Similar considerations apply to examples of brains in vats. See note 14 above.

rejected is the idea that an agent's believing *S*—even if the agent is a language user among language users—entails that the agent performs or can perform an appropriate speech act of assent. There may be other markers of belief. We have also suggested that there are cases where sincere assent to a sentence even on the part of a competent, reflective language user need not be a sufficient condition, an overriding guarantee of believing, since it denies that a person's nonverbal actions that seem to run counter to his avowals can be evidence against his having that avowed belief. Briefly stated, we have questioned a principle generally accepted as noncontroversial;

> Q: *x*'s assent to a sentence '*S*' *entails* that *x* believes that *S*, where the conditions on *x* of sincerity, linguistic competence, and reflectiveness obtain.

The widespread acceptance of the disquotation principle Q is not wholly without ground. My suggestion is that, on a broader view, other actions might belie the agent's words, and sincere assent might not be the privileged marker of believing. Let us suppose, however, for the discussion that follows, that our agent's other actions *are* coherent with his sincere assentings and he makes no logical mistakes. If that is the case, then, on principle Q, his sincere assenting to '*S*' does go over into a belief, for it is assumed that a linguistically competent, reflective speaker can reliably report what he believes. Where conditions including coherence hold, it seems that there still remain, on the language-centered view, some puzzles that can be resolved where Q is viewed as falling under D, i.e., where in assenting to a sentence under the appropriate conditions an agent is acting as if a state of affairs described by that sentence obtains.

Q and a Puzzle about Beliefs in Nonexistence

In "Speaking about Nothing" Keith Donnellan[17] recounts the actual case of *The Horn Papers,* which were purported to be the published diaries of one Jacob Horn, a colonial American, and were so viewed until historians disclosed that the *Papers* were a hoax. Consider someone, call her "Sally," who, having read *The Horn Papers* and being unaware that they were part of a fabrication, sincerely assents to 'Jacob Horn lived in Washington County, Pennsylvania'. The syntactical

---

17. *Philosophical Review,* LXXXIII (1974): 3–30. *The Horn Papers* was launched as history; hence the characterization "hoax" rather than "fiction."

proper name 'Jacob Horn' is not a genuine referring name but an invention of the hoaxster. Elaborating on Donnellan's views, the act of assenting to what seems a perfectly formed sentence 'Jacob Horn lived in Washington County, Pennsylvania' should not carry over into a belief. Backtracking her acquisition of the name does not terminate in a person named 'Jacob Horn'. But Sally need not know that. It is simply that the purported state of affairs described is not a complete state of affairs. It is as if she had assented to 'z lived in Washington County, Pennsylvania' where 'z' is a variable. What she assented to does not describe a closed structure. Of course, Washington County, Pennsylvania, *is* a constituent, and lived in *is* a relation; the partial structure does not wholly lack constituents, but it lacks a constituent needed to make it count as a state of affairs. Yet Sally's was a sincere assent to 'Jacob Horn lived in Washington County, Pennsylvania'. She is competent, not conceptually confused, and reflective. We have also assumed that she makes no logical errors and is broadly rational. She just lacks the relevant knowledge that 'Jacob Horn' is without a referent. In such a case, on disclosure that the syntactical name 'Jacob Horn' does not refer, Sally should say, on my proposed analysis, that she only *claimed* to believe that Jacob Horn was a resident of Washington County, Pennsylvania, in the first instance. Of course the disquotation principle simply has as antecedent 'x assents to 'S' ', and it may be an implicit assumption of the disquotation principle Q that linguistic competence and absence of conceptual confusion will rule out assent to a sentence that is not fully interpreted. But, as in the present example, that is unjustified. A rational competent agent can, on the psychological side of the believing relation, appear to "hold" a sentence "true" that lacks a truth value altogether. Nor will interjecting Jacob Horn as a possible person work, for reasons discussed elsewhere.[18] Suffice it to say that Sally took 'Jacob Horn lived in Washington County, Pennsylvania' to be making a historical claim, as the author of *The Horn Papers* intended.

Of course, in the context of the example, Sally may assent to related sentences that *are* fully interpreted, such as 'There was a person named "Jacob Horn" born in Washington, Pennsylvania, who kept a diary', which will carry over into a belief. Her nonverbal behavior may also be affected in predictable ways. She may engage in what she believes to be historical research, such as searching out "facts" not narrated in *The Horn Papers* about the purported person, as she

---

18. "Dispensing with Possibilia," *Proceedings of the American Philosophical Association*, XLIX (1976): 39–51. Also, "Possibilia and Possible Worlds," this volume.

might have done had there been such a person. On the present object-centered view, her behavior is rational and explicable, although she could not have believed that Jacob Horn was born in Washington, Pennsylvania. Principle Q requires an additional condition: that '*S*' be a fully interpreted sentence.

In the case of failed reference, as in the above example, what goes wrong is that the troublesome sentences *appear* to have a "content" that they do not have. In discourse that purports to be about our actual world it is presupposed that proper names, when used as in the above example, do refer.[19] Sentences with such reference failures mislead. In such cases we should disclaim having had a belief, despite sincere assenting.

It should be noted that this case of reference failure does not present a problem for a Davidsonian language-centered view of Q, since Davidson requires that '*S*' be fully interpreted, and, in the absence of a referent of 'Jacob Horn', the sentence 'Jacob Horn was born in Washington, Pennsylvania' lacks a complete interpretation despite appearances and is not a truth bearer.

Q and a Puzzle about Identity and Contradiction

Although for the purposes of discussion of puzzles we have assumed that the epistemological agent is rational and a faultless logician, it *appears* to follow from Q on a language-centered view that such agents can come to "believe" a contradiction. What I want to argue is that, under D and the object-centered view, that does not follow. At worst, what may occur is that a rational, logical agent may be in a believing relation to an impossible state of affairs.

Consider the following example.[20] Someone, call her "Sally," is

19. In "Modalities and Intensional Languages," this volume, such a directly referential view of proper names was proposed. There, I say, "To give a thing a proper name is different from giving a unique description. This [identifying] tag, a proper name, has no meaning [as contrasted with having reference]. It simply tags. It is not strongly equatable with any of the singular descriptions of the thing." It should be noted that, on this view, proper names are *not* assimilated to descriptions, even "rigid" descriptions. Kripke, in "Naming and Necessity," in *Semantics of Natural Language*, ed. G. Harman and D. Davidson (Dordrecht: Reidel, 1972), classifies proper names as "rigid designators" along with rigid descriptions, thereby obscuring the difference in semantical relationship between a proper name and the object named as compared with the relationship between a rigid description and the object described. Kripke in "Discussion," present volume, interpreted my views as including the position that "the tags are the essential denoting phrases for individuals." That was not part of my account, but we can see in those 1961 remarks Kripke's move toward his theory of "rigid designators."

20. See Kripke, "A Puzzle about Belief." My example is an analogue of Kripke's case of Pierre's coming to believe a "contradiction."

rational in the wide sense and an impeccable logician. She also succeeds in maintaining in her actions, including her speech acts, a norm of coherence. In such a case we can normally take her sincere assentings as privileged markers of believing.

In the 1930s Sally became acquainted with Alexis Saint-Léger of the French Foreign Office. On the basis of information available to her she assents to the sentence 'Alexis Saint-Léger is not a poet'. Some years later Sally meets a poet, St.-Jean Perse, at the United States Library of Congress. Time has not been kind. St.-Jean Perse is not recognizably Alexis Saint-Léger, and Sally assents to the sentence 'St.-Jean Perse is a poet'. Since unbeknownst to her 'Alexis Saint-Léger' and 'St.-Jean Perse' name the same person, he *is* the constituent in the states of affairs encoded into or described by the sentences 'Alexis Saint-Léger is not a poet' and 'St.-Jean Perse is a poet'. Each of those sentences taken separately describes a possible state of affairs. Saint-Léger might not have written poetry. Circumstances could have prevented that, as well as his serving in the French Foreign Office. Given that Sally is logical, she will also assent to the sentence 'Alexis Saint-Léger is not a poet and St.-Jean Perse is a poet', which unbeknownst to her describes an impossible state of affairs. Given that she knows about the identicals having to have all properties in common, including uniqueness, she assents to 'Alexis Saint-Léger is not St.-Jean Perse', which on the necessity of identity also describes an impossible state of affairs. She would of course not assent to 'Alexis Saint-Léger is not Alexis Saint-Léger', which unbeknownst to her describes the same impossible state of affairs as 'Alexis Saint-Léger is not St.-Jean Perse'. She would also assent to ' 'Alexis Saint-Léger' and 'St.-Jean Perse' name different persons', but that sentence is not descriptive of an impossible state of affairs. Our background theory requires only that names, once *given,* retain a fixed value.

This puzzle is irksome to those who hold the language-centered view of belief in which the objects of believing are either sentences or those propositions that mimic sentences in having properties like true, false, contradictory, valid, and the like. On that account, given the disquotation principle, Sally's assentings to 'Alexis Saint-Léger is not identical to St.-Jean Perse' and 'Alexis Saint-Léger is not a poet and St.-Jean Perse is a poet', go over into beliefs. But on the language- or proposition-centered views, the objects of believing are sentences or propositions. Since, semantically speaking and unbeknownst to her, the same person is assigned to both names, the assentings go over into *contradictory* sentences or propositions; so she seems to believe contradictions. Yet she has justification for arriving at her beliefs and has made no logical errors. She lacks other information. Why should the mere lack of information lead one to believe contradictions although one began with only seemingly justifiable albeit some false premises?

The language-centered theorist is baffled. What *does* Sally believe, he asks? If she believed that Alexis Saint-Léger is not a poet, then that is the same *proposition* as that St.-Jean Perse is not a poet. Does she or does she not believe that St.-Jean Perse is a poet?

But note how differently our presently proposed view accommodates the language-centered theorist's puzzle. Sally was introduced to Alexis Saint-Léger in the French Foreign Office. The individual in the state of affairs described by the sentence she assents to, i.e., 'Alexis Saint-Léger is not a poet' is that person: not the essence of Alexis Saint-Léger, not the concept Alexis Saint-Léger, not the sense of the name 'Alexis Saint-Léger'.

When later she meets St.-Jean Perse, he, that person, is a constituent in the state of affairs described by the sentence 'St.-Jean Perse is a poet'. What Sally lacks is information that would permit her to *reidentify* Saint-Léger. The sentences she assents to are not surface contradictions. The syntactical *form* of some of the sentences she assents to that describe impossibilities do not have the surface form of logical falsehoods. It is surely possible for different syntactical names to name the same thing.

Sally acquired the names 'Alexis Saint-Léger' and 'St.-Jean Perse' on two different occasions and under different circumstances, as very likely did Saint-Léger. The chain of communication in the public language will carry the second name into the first and finally to the object named. The value of those names, if they refer, is fixed, but not by some known set of identifying descriptions that permitted a determination by the agent. As a practical matter, even a chain of communication may be practically irretrievable. Unlike the evidence in the Jacob Horn case or the Saint-Léger/Perse case, historical records may not be available sufficient to make a determination of reference or absence of reference. It is such situations that prompt research (often frustrated) into, for example, the historical Homer or the historical Robin Hood.

On our proposed view of believing, x believes that S when x has a disposition to act as if a certain state of affairs obtained. A sincere assenting to 'S' by a rational, logically impeccable, but nonomniscient agent can often serve as a privileged marker of believing that S. But, unlike linguistic propositions, states of affairs obtain or do not obtain, must obtain or cannot obtain. They are not true or false, contradictory, valid or invalid. Some of the sentences Sally unwittingly assents to would seem to lead her to be in a believing relation to an impossible state of affairs, but, though she *assents* to a sentence translatable into a logically false sentence, she doesn't *believe* a contradiction, as demanded by the language-centered account.

Note that, on the dispositional account given by principle D, agent-centered circumstances as well as external circumstances are conditions on x's being disposed to act as if S, that state of affairs, obtained. There are (1) circumstances under which Sally is disposed to act as if Alexis Saint-Léger (i.e., St.-Jean Perse) is not a poet and (2) circumstances under which she is disposed to act as if Alexis Saint-Léger (i.e., St.-Jean Perse) is a poet. Indeed, among those circumstances in (1) are those in which she also assents to 'Alexis Saint-Léger is not a poet' and does not assent to 'St.-Jean Perse is not a poet'. Similarly for the circumstances in (2) under which she assents to 'St.-Jean Perse is a poet' and does not assent to 'St.-Jean Perse is not a poet'. Note that assenting is a speech act that occurs at a time and place and *under circumstances*. Under those variant circumstances her assents, in accordance with Q, each go over into a belief. Indeed on Q, the above assents *each* carry over into a believing relation to a possible state of affairs.

But Sally will also assent in some circumstances to 'St.-Jean Perse is a poet and Alexis Saint-Léger is not a poet' and, on Q, that would put her into a believing relation to an impossible state of affairs. In assenting to a sentence that, given still unknown reidentifications, might come, on logical grounds but unbeknownst to her, to have on substitution the form of a surface contradiction, she is on D disposed to act as if an impossible state of affairs obtained.

On such a theory, and given Q, the answer to the question of what Sally believes about Alexis Saint-Léger and St.-Jean Perse is straightforward. If, as we have proposed, the objects of believing are not linguistic entities, then, on the disquotation principle, we can say that she believes what she says she believes: that Saint-Léger is not a poet and that St.-Jean Perse is a poet, and, given that she will assent to logical consequences of sentences assented to, she may also act as if some impossible states of affairs obtained, to the extent that such actions are describable. Her assents, her logical reasoning, and her evidential grounds are *not* incompatible with having such dispositions. She is not omniscient. So, seeing no incompatibility between rationality in the narrow sense and believing impossibilities, should we not let the matter rest? A puzzle has been solved. Furthermore, there seems to be ample other evidence that impossibilities are believed.

A Controversial Proposal

Given the present account so far, we are led not to the puzzling conclusion that a logically rational agent "believes a contradiction" but
only to the conclusion that under certain circumstances she is in the
believing relation to an impossible state of affairs, held to be describable by a sentence to which she assents. A puzzle has been
solved. Nevertheless I should like to propose, on other considerations,
a modification of Q, which disallows believing impossibilities. On that
proposal, an agent can *claim* to have such beliefs but will be mistaken
in so claiming. Just as norms of truth lead to retroactively revisable
knowledge claims, norms of rationality should lead to retroactively
revisable belief claims. What is being proposed is that, whatever the
psychological dimension of the belief state, on disclosure of impossibility that belief *claim* should be viewed as mistaken.

Despite prevailing views to the contrary, the thesis that one cannot
believe impossibilities has its advocates. Berkeley[21] argues that to
believe propositions that entail contradictions is illusory. He says of
such propositions that "they are an instance wherein men impose upon
themselves by *imagining* they believe those propositions."

There is, of course, agreement that a rational agent does not *assent*
to simple formal surface contradictions such as '*S* and not *S*', but in
such a case of assenting to an overtly and formally contradictory sentence one can claim that a condition on the disquotation principle had
not been met. The agent is said to be *conceptually confused*. He has
not grasped the meaning, and he does not comprehend the semantics
of words like 'not', 'and', and so on.

Cases of deductive failure that lead to assenting to sentences that
are equivalent on substitution to contradictions can sometimes be attributed to mistakes in calculation. But, among those like Berkeley,
Wittgenstein, and some positivists who rejected belief in impossibilities that are represented by logically or formally false sentences, there
was an underlying argument for rejecting such beliefs that would seem
to apply to all cases of impossibility, including those that have their
origin, as in Sally's case, in deficient information.

Wittgenstein[22] is concerned in the *Tractatus* with those necessities
and impossibilities that are given by tautologies and contradictory
propositions of *whatever* complexity. He argues that a *significant* prop-

21. *Principles of Human Knowledge,* ed. C. M. Turbayne (Indianapolis: Bobbs-Merrill, 1970), p. 273.

22. *Tractatus Logico-Philosophicus* (London: Kegan Paul, 1922). See especially
4.461–4.466, 5.1362, 5.142, 5.43a, 6.11.

osition has to describe a definite situation such that the situation may or may not obtain. A proposition must admit an alternative truth value. Where a proposition does not admit of alternative values, i.e., is not contingent, he says it lacks sense, although he adds that it isn't exactly nonsense either. "Tautologies and contradictions . . . do not represent possible situations." "The truth of a tautology is certain, of contradiction impossible," and therefore instances of both lack significance. If 'S' is senseless, then 'x knows that S' is senseless even where 'S' is a tautology. On the same ground he should say that 'x believes S' where 'S' is a contradiction is also senseless. But if what informs his argument is that it is the impossibility of a proposition S being false or the impossibility of a proposition S being true that makes it an improper object of a propositional attitude, then the *origin* of those attitudes should not matter. A false identity claim, for example, is necessarily false, never mind how it was arrived at.

A later Wittgenstein[23] softened somewhat toward the range of significant propositions. Necessary truths and falsehoods are no longer denied sense. But he does hint at a distinction between believing and claiming to believe impossibilities. He says, "I feel a temptation to say one can't believe $13 \times 13 = 196$. . . . But at any rate I can *say* 'I believe it' and act accordingly" (italics mine). That hint, if it is a hint, can be elaborated into a proposal. Present received views insist that if we say of an agent that he knows that S and S turns out not to be the case, we alter our ascription. This can be done retroactively. Disclosure of the falsity of 'S' would falsify the knowledge claims of all who ever claimed they knew that S. But it would seem that we do not likewise retroactively disclaim belief in impossibilities—unless we take something like Wittgenstein's early radical view that impossible states of affairs are not states of affairs at all and hence not proper objects of epistemological attitudes. There were after all those who thought that angles could be trisected with a compass and a ruler, or that the consistency of arithmetic was provable on some canonical criterion of proof, or that Shakespeare was the Earl of Oxford. Those propositions are necessarily false, yet what is claimed seems to have been believed. But I do see an advantage to a revision of the disquotation principle; not Wittgenstein's early radical proposal (for tautologies and contradictions are surely meaningful) but a proposal that allows a distinction between believing and claiming to believe. We propose a modification of Q as follows. In addition to linguistic com-

---

23. Remarks on the *Foundations of Mathematics* (Oxford: Blackwell, 1956): I-106, p. 31.

petence, sincerity, and reflectiveness of the agent, we add the condition that her actions, including her speech acts, are coherent and preserve a norm of rationality in the wide sense. The additional condition is required if assenting is to count as an *overriding* marker of belief.

> Q': (1) x assents to 'S', (2) 'S' is a fully interpreted sentence in x's language, and (3) S is possible, together entail that x believes that S, where conditions of sincerity, competence, and reflectiveness obtain.

If conditions (2) and (3) are not met, x's assenting does not carry over into a belief. The two puzzling cases are accommodated. In *The Horn Papers* example, (2), the condition of complete interpretation of 'S', is not met. Nevertheless a rational account can be given of why Sally *claimed* to have a belief that Jacob Horn lived in Washington County, Pennsylvania, despite the absence of a fully structured state of affairs that is a proper object of believing. In the Alexis Saint-Léger/St.-Jean Perse example the state of affairs is properly constituted. Here the disclosure of the truth of an identity sentence, e.g., 'St.-Jean Perse is identical to Alexis Saint-Léger' would reveal the logical falsehood of some sentences to which Sally assents. In such a case Sally might say that she only claimed to believe that Alexis Saint-Léger was not the same as St.-Jean Perse, for such a belief comes to believing of a thing that it is not the same as itself and that does not meet logical norms of rationality. Just as the falsehood of 'P' excludes knowing that P, the necessary falsehood of 'P' excludes believing that P, whatever the agent's knowledge claims or belief claims, respectively.

The revisionary proposal places conditions on when a *speech* act of assent goes over into a belief. It does not place a possibility condition on the beliefs of non-language-using agents. The condition of possibility is grounded in norms of rationality just as, with respect to knowledge, the condition of actuality is grounded in norms of truth. Such norms would seem to require second-order conceptualization and reflection open only to language users.[24]

If believing S is minimally a disposition to act as if a certain state of affairs obtained and if such a state of affairs could not possibly obtain, a rational language-using epistemological agent is in a position to ask what would count as acting as if it did obtain. Many actions would be rendered incoherent, many ends frustrated. If the speech act

---

24. This is what seems to be at the center of Davidson's view mentioned at the outset of this paper.

of assenting is one of those actions that mark our believing, then we would be acting as if '*S*' were true. But what sense can we make of "acting as if '*S*' were true" when either '*S*' has no truth value or '*S*' is necessarily false? Of course, on the revisionary proposal, we could not ascribe belief to those who claimed to believe that the Fountain of Youth is in Florida (where 'Fountain of Youth' is mistakenly taken to be a directly referring name) or that an angle can be trisected with compass and straight edge or that Shakespeare is identical with the Earl of Oxford. We could say only that they claimed to have these beliefs. They were not irrational agents, and we can explain why they made those claims. This revision does not do too much violence to plausible usage, for these agents did have other proper beliefs that explained (1) their assent to sentences that described impossible states of affairs, (2) their assent to sentences that, appearances to the contrary, are not fully interpreted.[25]

Some Concluding Remarks

It must be emphasized that the account of belief given by (D) may be detached from the proposed controversial constraint (Q' versus Q) on the disquotation principle. However, D and the principle Q' do have radical consequences for the semantics of belief sentences and belief reports, affecting such questions as whether the belief "operator" can be factored out of or distributed over a conjunction, and the like.

Also and more radically, this view suggests that the "sentential" or "quasi-sentential" or "propositional" account of epistemological attitudes and mental representations is inadequate and must be replaced by some expanded account of mental states and information processing, which does not take sentences or analogues of sentences as the objects of believing.

25. See "Possibilia and Possible Worlds," this volume.

Index